JOURNAL OF CONASENSE

Communication
Navigation
Sensing
Services

Volume 1, No. 2 (May 2014)

Special Issue on
*Interaction of Communications, Navigations and Sensing with Control
and Automation for Smart Services Provision*

Guest Editors
Ernestina Cianca
Mauro De Sanctis

JOURNAL OF CONASENSE

Aim

The overall aim of the CONASENSE Journal is to provide a common platform for exchanging ideas among the communities, both the academic and industrial, involved in the fields of Communications, Navigation and Sensing, with emphasis on multidisciplinary views and Smart/Intelligent services that require the effective and efficient integration of these three fields of research and development.

Scope

The Journal will publish articles on novel research and the latest advances, in the field of communication (in particular, wireless communication), navigation and sensing with special emphasis on the challenges, new concepts and future enablers for the interaction/integration of these technologies for the successful provision of i smart/intelligent services.

The fields of interest include:

- All communications/sensing/navigation systems and techniques, protocols which enable awareness of the physical environment, effective and fast feedback loops between actuation and sensing, a flexible and cognitive architecture which comply with essential requirements like safety, security, near-zero power consumption as well as size, usability and adaptability constraints.
- Control theory aspects in presence of wireless or lossy feedback links (i.e. network control theory), distributed control systems;
- Services and applications such as smart grid, Ambient Assisted Living, Ambient-Intelligence, Smart Cities, Smart Environment, Context-aware services, location-based services, e-Health, but more in general innovative services and applications for contributing to solving societal challenges.
- Data management such as data mining, data retrieval, decision-making algorithms.

Published, sold and distributed by:
River Publishers
Niels Jernes Vej 10
9220 Aalborg Ø
Denmark

Tel.: +45369953197
www.riverpublishers.com

Journal of CONASENSE (Communications, Navigation, Sensing and Services)
is published three times a year.
Publication programme, 2014: Volume 1 (3 issues)

ISSN: 2246-2120 (Print Version)
ISSN: 2246-2139 (Online Version)
ISBN: 978-87-93237-17-9 (this issue)

JOURNAL OF CONASENSE

Volume 1, No. 2 (May 2014)

Editorial Foreword

"Interaction of Communications, Navigations and Sensing with Control and Automation for Smart Services Provision"

Control is a key element to provide "Smart" services as "Smart" means the ability to react and adapt to events.

This special issue of the CONASESE journal collects contributions on ICT architectures and key ICT issues to implement control and automation through a large number of smart objects pervasively deployed in the physical world and wirelessly connected and equipped with some control logic running at the core of the application.

Therefore, the power grid becomes "smart" when it is able to react and adapt to any event occurring in any point of the grid (generation, transmission, distribution, consumption) by adopting proper strategies and countermeasures. To make this a reality, a bi-directional flow of information is needed among the different actors of the grid at different points of the grid. Many of the elements of a Smart Grid implementation are already available, including smart meters, automated monitoring systems and power management systems. The communication network has a crucial role and its cost and performance will greatly impact utilities' revenues and the capability of the new grid to meet its ambitious objectives. For sure, the communication network will be heterogeneous, including wired and wireless, as well as private and public solutions. In this framework, it is challenging to enable smart grid operations (automation and control), over these imperfect, heterogeneous networks. Paper #1 "Adaptive Monitoring and Control Architectures for Power Distribution Grids over Heterogeneous ICT Networks", based on the activity of the EU SmartC2Net project, provides an overview of the ICT architecture for smart grid operations. In particular, the ICT architecture for the External Generation Site is detailed and its evaluation approach is presented.

Tightly related and partially overlapped with the concept of SmartGrid is the concept of Smart Building. For instance, the smart metering is an element of the smart grid as well as the smart building. From the grid side, it enables more effective demand side management techniques, and from the user point of view, it enables the control of the energy consumption at appliance/device level. Again, control is crucial to make smarter buildings. In this framework, an important element is the Building Management System (BMS) which is a computer-based control system installed in buildings that controls and monitors the building's mechanical and electrical equipment such as ventilation, lighting, power systems, fire systems, and security systems. Current BMS lack of personalization and service adaptation. Paper "Applications of Machine Learning and Service Oriented Architectures for the new Era of Smart Living" presents the enhancements of a proprietary BMS product, namely Ecosystem, which are needed to address high demand for personalization and service adaptation in the new era of ICT.

Moreover, the M2M paradigm is the key enabler of any secure and dependable automation of tasks needed to implement smart services. This automation is realized by a closed loop between data sensing and actuation, and implemented through a large number of smart objects pervasively deployed in the physical world which communicate with each other. In critical scenarios, however, such as public safety systems or personal health-care, the control is only effective if it is able to operate while meeting specific real-time requirements. In this framework, paper "A framework for Quality of Service support in Things as a Service oriented architectures" presents a general solution to support the broad range of QoS requirements that characterize heterogeneous M2M applications. The proposed solution is designed with particular attention to scalability and reliability in order to be integrated into large Internet-of-Things (IoT) deployments, and comprises a QoS model and a framework for QoS negotiation and resource allocation.

Finally, security is a well known challenge for IoT scenarios. Embedded IoT devices, with network connectivity but without displays or keyboards, are a challenge for security. How can the end-user securely connect a washing machine or lightning system to a previously unknown cloud service or remote controller when there are no means to input passwords? Paper "Mediated Security Pairing for the Internet of Things" presents challenges and approaches for security pairing interface-restricted globally distributed things. In particular, Authors presents an analysis on how user-friendly security establishment approaches – out-of-band and unauthenticated location-based pairing – can be applied in situations where counterparty is far away or has incompatible interfaces.

Editors would like to thanks the Authors that contributed to this Special Issue. The above papers provide just some examples on ICT issues to be faced to provide effective control, which is crucial to implement smart services and we hope that this Special Issue will stimulate more focused "interdisciplinary" research on the interaction between ICT, in its wider sense (Com/Nav/Sensing), and control.

Ernestina Cianca
Mauro De Sanctis
University of Tor Vergata, Rome, Italy

A Framework for Quality of Service Support in Things-as-a-Service Oriented Architectures

E. Mingozzi, G. Tanganelli and C. Vallati

Dip. Ingegneria dell'Informazione, University of Pisa, L.go L. Lazzarino 1, I-56122, Pisa, Italy

Received 28 May 2014; Accepted 28 June 2014
Publication 31 August 2014

Abstract

M2M applications allow the secure and dependable automation of tasks to improve industrial productivity and quality of life of citizens. Task automation is realized by a closed loop between data sensing and actuation, and implemented by means of a large number of smart objects pervasively deployed in the physical world, and control logic running at the core of the M2M application. In critical scenarios, however, such as public safety systems or personal health-care, the control is only effective if it is able to operate while meeting specific real-time requirements. In this work we present a general solution to support the broad range of QoS requirements that characterize heterogeneous M2M applications. The proposed solution is designed with particular attention to scalability and reliability in order to be integrated into large IoT deployments, and comprises a QoS model and a framework for QoS negotiation and resource allocation. The framework has been included in the runtime platform for M2M applications developed by the EU-FP7 project BETaaS: Building the Environment for the Things as a Service.

Keywords: Quality of Service; M2M applications; Internet of Things; Service Level Agreement; SLA Negotiation.

Journal of Communication, Navigation, Sensing and Services, Vol. 1, 105–128.
doi: 10.13052/jconasense2246-2120.121

1 Introduction

The Internet of Things vision foresees a seamless integration of things from the physical word into the Internet. Things now working in isolation will be accessible through the Internet enabling a new generation of applications whose only boundary is imagination [1]. The shift from a set of vertical systems working in isolation to a horizontal platform integrating different physical objects from heterogeneous environments is recognized as the most important innovation to create a fertile soil for the development of new applications. BETaaS[1] – Building the Environment for the Things as a Service – is a European project funded under the Seventh Framework Programme, which aims at designing a runtime platform for Machine-to-Machine (M2M) applications. Built on a distributed architecture made of a local cloud of gateways, the BETaaS platform transparently integrates existing heterogeneous M2M systems. Smart things are exposed through a service-oriented unified interface, the Things as a Service model, which allows applications to access things transparently regardless of the 'physical layer' technology and the location.

Built-in support for non-functional requirements such as reliability, scalability, energy efficiency and quality of service (QoS) are included in the architecture design. Among them, QoS is of particular importance since it has been identified as a key non-functional requirement to enable many IoT-related application scenarios, and M2M ones in particular [2, 3]. Security and emergency content in smart city and home environments often has strict real-time requirements. As an example, latency (besides dependability) is a critical factor for applications such as real-time sensor monitoring in personal health-care or public safety systems [4]. On the other hand, other applications like, e.g., road traffic management applications for urban mobility, though less sensitive to delay bounds, may nevertheless benefit from receiving some form of soft real-time treatment, at least for a subset of their provided services (e.g., 'urgent' alert notifications) [5]. Moreover, applications involving streaming of multimedia content like video surveillance that consumes high bandwidth will require full support from the platform not only to guarantee an acceptable level of service but also to avoid saturation and waste of network resources.

In this context, efficient support for heterogeneous QoS requirements is a non-trivial challenge. The number of shared communication and computational resources that will be highly heterogeneous and constrained in a variety

---

[1]http://www.betaas.eu/

of manners represent a complex field to operate: wireless access networks will be characterized by limited and time-varying throughput and error rates; services in the physical world may be provided by devices which could be mobile – thus not always available – and energy-constrained. In order to handle this heterogeneity, a comprehensive QoS management framework is embedded at any level of the BETaaS platform. The framework operates at different time scales, i.e., resource provisioning and runtime, and comprises a general QoS model, SLA-based negotiation and admission control, and optimized resource provisioning.

In this paper we present the QoS framework adopted in the BETaaS platform. The architecture has been designed to fit within the BETaaS architecture described in [14]; however, it is important to highlight that the proposed framework can be adopted without radical modifications also in other architectures. The proposed architecture is designed with the following objectives: (i) guaranteeing scalability over large IoT deployments charac- terized by heterogeneous devices, (ii) supporting M2M applications with a vide range of requirements, (iii) exploiting the large number of smart objects that are expected to be connected providing equivalent services (equiva- lent things) in order to allow the definition of efficient resource allocation algorithms.

To this aim, a distributed design is adopted in order to guarantee system scalability, a critical requirement in large-scale IoT deployments. To avoid data inconsistencies, race conditions, and long response times that can affect functionalities implemented in a distributed manner over large deployments, critical system functions (e.g., admission control) are provided through a centralized point of decision, which however can be implemented in a distributed manner for scalability and resiliency over a subset of nodes. Support for heterogeneous applications is guaranteed through Service Level Agreements that allow the negotiation of the desired QoS including also hard real-time requirements. A SLA negotiation framework is included in the design to expose to applications an interface to negotiate the desired QoS level.

The rest of the paper is organized as follows: in Section 2 an overview of existing solutions to provide QoS for IoT is provided, in Section 3 a short description of the BETaaS architecture is given. Section 4 presents the QoS negotiation framework included in the platform as interface to support SLA negotiation, while Section 5 illustrates the architecture of the QoS framework included in the platform to enforce and monitor QoS levels. Eventually conclusions are drawn in Section 6.

2 State of the Art

Quality of Service support has been identified as a key non-functional requirement to enable many IoT-related application scenarios. Although several proposals have been presented in order to enable end-to-end QoS in constrained environments, all these works focus on a particular technology or address a specific sub-problem and do not propose a solution to handle the problem entirely. An example is the field of Wireless Sensor Networks (WSN) in which many models for QoS have been developed for the MAC layer [15, 16] or the routing protocol [17, 18]. The future IoT world, however, will go beyond current WSNs with a new generation of devices such as actuators or smart cameras that differ significantly from traditional sensors.

Several proposals aimed at introducing QoS support have been proposed for Service Oriented Architectures (SOA) where QoS is a crucial requirement. A first QoS framework for SOA is presented in [20], where applications request services with certain requirements and the framework manages to find a possible allocation, among the set of registered service, in order to fulfill QoS requirements. In [19] authors propose a QoS framework supporting also Real-Time requirements in a distributed heterogeneous environment. Approaches defined in the context of SOA systems lack of support for constrained devices that introduce new non-trivial issues and would require significant modifications.

To the best of our knowledge, a first attempt to address such issues has been done in [21], where the use of web services for sensors integration is proposed. However, this approach aims at implementing a service-oriented middleware directly on the nodes, which is not always feasible due to their constrained environment. To overcome these limitations, in [22] an adaptable middleware is proposed; the middleware functionalities can be configured to reduce their complexity in case of constrained devices such as sensors. The proposed solution exposes a SOA interface to applications in which a flexible QoS support is provided by means of Service Level Agreements (SLAs) between the applications and the middleware. The solution proposed, however, is specifically tailored to WSNs.

In the context of distributed real-time systems, in [23] a framework for dynamic resource allocation and re-distribution is presented, however it lacks of admission control functionalities that are important especially in constraint environments. To partially overcome this issue, authors of [24] propose a middleware that implements an admission control and a load balancer.

The latter in particular is responsible for optimizing resource allocation at run time by migrating tasks between processors, if necessary.

3 BETaaS Architecture

BETaaS goal is to build a content-centric platform for M2M applications which can transparently access things from different heterogeneous M2M systems. The platform is deployed on top of a distributed architecture made of a local cloud of gateways. Each physical object is connected to a local gateway; applications connect to one gateway which represents the entry point to access the rest of the platform. Applications can access physical objects through a service-oriented interface according to a Things as a Service model.

Seamless integration is achieved through a layered architecture that includes three main layers (see Figure 1): the *Service layer*, the *Logical layer*, and the *Physical layer*, respectively. The logical layer is further subdivided into two sub-layers, i.e., the *Things-as-a-Service (TaaS* for short*)* layer and the *Adaptation* layer. This structure allows the integration of different existing M2M systems, the so-called underlying 'physical layers'. Each M2M system is integrated into the platform through a specific Adaptation layer, which conforms the access to a unified interface.

Core of the overall architecture is the TaaS layer. Each gateway must implement only one TaaS instance called TaaS local component. The set of TaaS local components interacts with each other to form a distributed TaaS layer. TaaS local components leverage on the functionalities offered by the

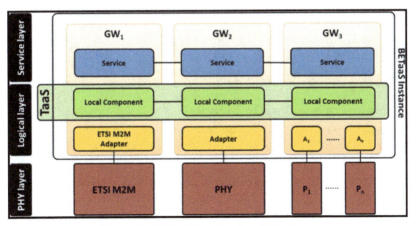

Figure 1 BETaaS functional model instance

local Adaptation Layers in order to access the smart objects. For each local smart object a corresponding *Thing Service* is created. A Thing Service is the representation of a single thing inside the TaaS layer; this representation is used to expose the functionality of each thing to applications. In order to identify uniquely a thing service within the entire TaaS instance a thingServiceID is assigned to each Thing Service. Distinctive feature of TaaS is the ability to exploit the concept of *equivalent things*. Two things are considered equivalent if they can provide interchangeably the same Thing Service according to the context information associated[2]. When a group of equivalent things is discovered, the TaaS layer creates a new Thing Service that represents the generic service that can be provided by any one belonging to the group.

Top of this layered structure is the Service layer which provides services to applications leveraging on the things exposed by the TaaS layer. A service can either implement basic functionalities mapped directly to one single physical object, or extended functionalities exploiting several basic services. For a more detailed presentation of the BETaaS architecture and its TaaS model, we refer the interested reader to [14].

4 QoS Negotiation Framework

In order to handle the large variety of requirements that characterizes M2M applications, a QoS negotiation framework is included in the platform to allow applications to negotiate the required QoS.

4.1 QoS Application Models

In order to guarantee a QoS that can be negotiated by applications, a uniform QoS model has to be defined first. In the following, we first provide an overview of the state of the art on QoS models, and then we present the classification derived for the BETaaS platform.

4.1.1 Overview of existing QoS models

A QoS service model and relate requirements for M2M applications was developed in [6] and [7] with specific reference to cellular networks as access technology. In particular, authors of [7] group M2M applications into five categories: *mobile streaming, smart metering, regular monitoring, emergency*

---

[2]Methodologies to extrapolate and manage context information are out of scope of this work. For this reason, we assume that context information related with things is given as well as a methodology to identify equivalent things.

alerting, and *mobile POS* (Point Of Sales). Each category poses different QoS requirements to the underlying network:

- *Mobile streaming* traffic involves continuous video transmission at high data rate with lower priority than other traffic. On one hand, video requires high bandwidth, soft real-time delivery and low jitter. On the other hand, the traffic is error tolerant.
- *Smart metering* traffic is characterized by large sporadic burst of packets with a request-response pattern. Its priority is low and the transmission can be rejected in case of network congestion. Its transmission has to be reliable but the traffic is delay-tolerant.
- *Regular Monitoring* traffic is characterized by small periodic packets (the period is in the order of seconds). Its priority is low and it does not have real-time requirements. Transmission reliability, instead, is critical.
- *Emergency Alerting* traffic is the most critical category. It is characterized by bursts of data which require the highest priority. The packet size is not predictable and a real-time and reliable transmission is needed.
- *Mobile POS* traffic is characterized by bursts of data with low priority. Real-time transmission is not required but reliability is a crucial requirement.

Starting from this traffic categorization, the authors uniform QoS requirements in order to cover both H2H and M2M services. The categorization is based on the main features of three types of services: *conversational, data transferring* and *emergency alarming*, characterized respectively by *real-time transmission, data accuracy* and *transmission priority*. Based on this three macro types seven service categorizations ranging from real-time to best-effort service have been developed.

In [8] three service models are defined based on the following factors: *interactivity, delay* and *criticality: Open Service Model*, interactive, non real-time and non mission-critical; *Supple Service Model*, interactive or non-interactive (according to the user subscription), soft real-time and mission-critical; *Complete Service Model,* non interactive (continuous flow of data), hard/soft real-time and mission-critical.

Since BETaaS takes into account different heterogeneous scenarios it is important to consider a wide set of QoS requirements. However, M2M applications have their core functionalities relying on sensors and actuators which are constrained devices in terms of computational and communication capabilities. In the field of Wireless Sensor Networks (WSN) several studies have been carried on to highlight application QoS requirements.

In [9] a QoS management system with requirements designed specifically for WSNs is presented. In particular, energy consumption is considered as it is one of the most important QoS metrics for WSN. Different sensors, due to their own characteristics, have different requirements: battery powered sensors need a reduced data sampling frequency while sensors without energy limitations can use a higher data sampling frequency. However, in order to have a fine grained characterization of sensors, other capabilities are also taken into consideration besides energy operation. In particular, a sensor can be single or redundant. A single sensor can be polled by different sources at the same time; in this case the middleware has to manage the requests with a certain priority following the set of QoS metrics. A redundant sensor, instead, is a virtual sensor whose information is provided by a group of physical sensors which can provide an equivalent information. In this case, the middleware can issue the request to any physical sensor belonging to this group thus applying a load balancing policy. Another differentiation is related to sensor data: on one hand, multimedia sensors have strict delay requirements but are packet loss tolerant, on the other hand monitoring sensors do not have stringent real-time requirements but rely on reliable transmission. Eventually, the authors present a categorization based on sensor transmission type: *Event driven*, *Query driven*, and *Continuous*. Each category is mapped to different QoS requirements:

- *Event driven*: Medium access delay, Reliability, Energy consumption, Flexibility;
- *Query Driven*: Medium access delay, Reliability, Energy consumption, Flexibility;
- *Continuous*: Collision rate, Energy consumption, Interference/Concurrency.

In [10] the authors categorize applications in three major classes with different requirements: *inquiry tasks*, *control tasks* and *monitoring tasks*. Inquiry tasks require service timeliness and reliability. Monitoring tasks require reliability but the service is delay tolerant. A three-layer taxonomy is proposed: the Application and Service Layer which contains all services, the Network Layer which manages network functionalities and provides QoS support, and the Perception Layer, which is used for system monitoring. At the application layer, the QoS is intended by the user point of view and the focus is on Service Time, Service Delay, Service Accuracy, Service Load and Service Priority. At the network layer, QoS requirements are related with the network itself, but, in general, main indicators are: bandwidth, delay, packet loss rate and jitter. A method for passing QoS requirements from the upper layer (customers' requirements) to the lower layer (resource allocation and scheduling) is proposed. QoS communication and translation is another critical point: QoS requirements are different at each layer and depend on the corresponding level of abstraction. Authors partially overcome this problem by adopting a cross-layer approach and dividing services into four classes: Control, Guaranteed Service; Query, Guaranteed Service/Differentiated Service; Real-Time monitoring, Differentiated Services; Non Real-Time monitoring, Best Effort. Network Layer and Perception Layer use a QoS broker which is responsible for adapting QoS requirements received from the Application Layer.

4.1.2 BETaaS service classes

A significant trend emerges from the analysis of existing QoS models: the wide-range of application requirements is handled by means of a very detailed classification that results in numerous service classes. These approaches necessarily increase the complexity of the infrastructure without fully satisfying M2M applications which often require ad-hoc QoS assurances.

For this reason, in the BETaaS platform a simple schema composed by three classes of services has been adopted in order to reduce the complexity of platform management functionalities. At the same time, applications are allowed to customize their QoS requirements through a dynamic negotiation procedure. The three classes of services adopted in BETaaS are the following: *Real-time*, *Assured services* and *Best-effort*.

The *Real-time* class is designed for applications with hard response time requirements where timing responses are usually mission-critical, e.g., surveillance alarm system, healthcare monitoring, industrial control. The negotiation phase is based on parameters, such as response time or service period, expressed in a deterministic manner. The platform must strictly comply with the QoS guarantees provided to this class of applications.

The *Assured services* class instead is for applications with soft response time requirements. These applications usually tolerate some out-of-contract interaction; for this reason the negotiation procedure is based on probabilistic requirements. This class can be used by interactive application – ticketing or user information gathering – or may be used by tracking applications for logistic.

Finally, the *Best-effort* class is used by applications that do not require any guarantee such as an application for historical data collection.

4.2 QoS Negotiation Framework

The BETaaS platform has to allow applications to negotiate QoS through a standard protocol. The literature about services and, in general, Service Oriented Architecture (SOA) is rich and can provide standard solutions. In particular, a key feature in SOA systems is the service negotiation procedure. The WS-Agreement [11] and WS-Agreement-Negotiation [12] are the de-facto standards for SLA agreement negotiating, establishing and managing in the Web Service field. It is worth to mention that the WS-Agreement and the WS-Agreement Negotiation standards are already implemented by the WSAG4J [13] project. The implementation is Java-based, it is publicly available, and the code is well documented and stable.

The structure of the BETaaS QoS framework is illustrated in Figure 2. The layered architecture of the platform has been considered in the design: QoS functionalities are implemented at all layers, from the Service to the Adaptation, cooperating to achieve the overall QoS support in a cross-layer manner. QoS negotiation capabilities are provided to applications through a standard interface which can be implemented by means of the WS-Agreement-Negotiation protocol. In the BETaaS platform, to simplify the complexity of the implementation, applications specify the required services in a manifest that also contains the required QoS.

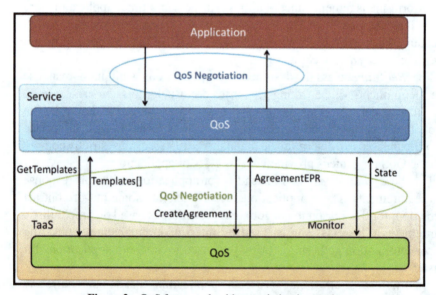

Figure 2   QoS framework with negotiation interactions

Applications are only the final consumers of services: each service can rely on more Thing Services simultaneously. Things expose their functionalities through Thing Services which are standard services from the application point of view. For this reason, a QoS negotiation procedure between the Service Layer and the TaaS Layer is also required. To this aim, the WS-Agreement-Negotiation protocol is adopted as the unified protocol for Thing Service SLA negotiation.

Two main advantages characterize the WS-Agreement-Negotiation protocol: flexibility and interoperability. QoS support provided by the BETaaS platform goes beyond the classic approach adopted in SOA architectures, i.e., QoS functionalities have to take into account the characteristics of the things and the unique requirements of M2M applications. To this aim, a flexible negotiation interface is needed to support future technologies with requirements that might not be already defined. On the other hand, the use of a standard protocol for QoS negotiation between the service and the TaaS layers assures interoperability by definition while guaranteeing also a higher level of implementation efficiency.

In detail, the WS-Agreement protocol defines the message exchange by two end-points in order to create a service level agreement – see Figure 2. In the first step, the service consumer requests all the available templates from the service provider. The consumer selects a template and creates a new agreement offer. Then, it sends the offer to the service provider to create a new agreement. The service provider replies with a confirmation or a rejection message, the confirmation contains an Agreement Endpoint Reference (AgreementEPR) generated to uniquely identify the committed agreement. An offer describes the service as well as the guarantees required for each service. Offers, templates and, in general, agreements are defined by an XML document with a specific schema outlined in Figure 3.

An agreement schema is composed by two main parts: Context and Terms Compositor. The Context provides information on both consumer and provider and optionally an expiration time that defines how long an agreement (and associated services) is valid. One or more Terms Compositor are specified to describe the services. The Terms Compositor is used to structure all the different terms related to each service by creating a so called Terms Tree. It is important to highlight that a Terms Compositor can contain zero or more other Terms Compositors. Each Terms Compositor is composed by Service Terms, a Service Term describes the service and gives a pointer to reference it and, finally, defines and evaluates the guarantees of the WS-Agreement. It is worth to note that multiple Service Terms can describe a single service in

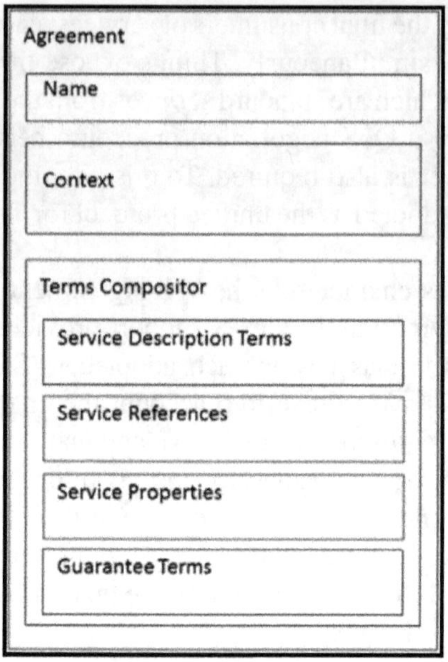

Figure 3   Agreement schema

an agreement. In fact, each Service Term describes a different aspect of the service.

The WS-Agreement Negotiation standard is an improvement built on top of the WS-Agreement. It gives to the consumer and provider, involved in the process of establishing an agreement, the capability of negotiating by means of offer (as the basic WS-Agreement) and counter offer.

In order to enable QoS negotiation within the BETaaS platform, specific templates are defined. As an example, Figure 4 shows the template included for Thing Service negotiation between Service layer and TaaS layer. This template compliant with the WS-Agreement specifications, defines a specific schema to describe the Service Description Terms required for the Thing Services invocation. The template follows the structure of the WS-Agreement standard: a Context section (<wsag:Context> tag) which contains the template name and the template id followed by a Service Description Term section (<wsag:ServiceDescriptionTerm> tag) which defines the terms of the Thing Service (<betaas:ThingService> tag). This section, in turn, includes the transaction ID needed to identify the Thing Service (<betaas:Definition> tag) and the QoS parameters, (<betaas:QoS> tag). In this example, a simple set of QoS parameters which can be included are defined:

```
<?xml version="1.0" encoding="UTF-8"?>
<wsag:Template wsag:TemplateId="1" xmlns:wsag="http://schemas.ggf.org/graap/2007/03/ws-
agreement">
  <wsag:Name>BETaaS-Template</wsag:Name>
    <wsag:Context>
      <wsag:ServiceProvider>AgreementResponder</wsag:ServiceProvider>
      <wsag:TemplateId>1</wsag:TemplateId>
      <wsag:TemplateName>BETaaS-Template</wsag:TemplateName>
    </wsag:Context>
    <wsag:Terms>
      <wsag:All>
        <wsag:ServiceDescriptionTerm wsag:Name="THING" wsag:ServiceName="THINGSERVICE">
          <betaas:ThingService xmlns:betaas="http://betaas.eu/schemas/betaas">
            <betaas:Definition>
              <betaas:transactionID>$TRANSACTIONID</betaas:transactionID>
            </betaas:Definition>
            <betaas:QoS>
              <betaas:MaxResponseTime>$MAXRESPONSETIME</betaas:MaxResponseTime>
              <betaas:MinAvailability>$MINAVAILABILITY</betaas:MinAvailability>
              <betaas:MaxRate>$MAXRATE</betaas:MaxRate>
            </betaas:QoS>
          </betaas:ThingService>
        </wsag:ServiceDescriptionTerm>
      </wsag:All>
    </wsag:Terms>
</wsag:Template>
```

Figure 4   Service layer - TaaS layer negotiation template

- *MaxResponseTime,* used to specify the Response Time of the Thing Service. In detail, it indicates the maximum delay between a Thing Service invocation and its response, measured at the service layer.
- *MinAvailability,* used to specify the availability of the Thing Service. This parameter is associated to each Thing Service and is, in principle, a static parameter. However, since the environment can change unpredictably the QoSMonitoring functionality must control and adjust this parameter continuously.
- *MaxRate,* is the maximum rate a Service can invoke the thing service, in other terms, a minimum inter-request time between two different requests from the same service.

5 QoS Framework

In order to enforce and monitor the negotiated QoS requirements, a pervasive QoS framework must be included in the platform architecture. Once the negotiation phase is performed, a SLA is established and applications could invoke thing services with the negotiated QoS level.

Before introducing more in detail the proposed solution, let us introduce the following assumptions:

- A thing service is uniquely identified by a thingServiceID (TSID).
- A Thing is uniquely identified by a thingID (TID).
- The AgreementEPR, retrieved during the negotiation phase, is used to authorize Thing Service invocation.

Specific requirement of the proposed platform is the exploitation of the possibilities offered by equivalent things. As a result of the integration of different systems, large IoT networks are expected to be characterized by a large number of equivalent things, which can potentially provide the same services. In this context, a QoS framework that considers equivalent things in its design can take full advantage of this variety to allow efficient management of resources.

In our QoS framework we propose a two-phase procedure, namely, *reservation* and *allocation*. The reservation phase is handled by a sub-component called *QoSBroker*. The QoSBroker manages the QoS negotiation, performs admission control and, most importantly, manages resource reservation by exploiting equivalent thing services. It also generates AEPRs to authorize thing service invocation. The allocation phase, instead, is managed by another sub-component called *QoSDispatcher*. The QoSDispatcher performs allocation of resources at time of thing service invocation. The QoSDispatcher can optimize the allocation by means of a number of parameters, e.g., in terms of energy efficiency. The reservation and allocation procedures are tightly connected. However, while the allocation procedure may be involved in each thing service invocation, the reservation procedure is executed only once at time of negotiation.

To avoid data inconsistencies, race conditions, and long response times, which can affect functionalities implemented in a distributed manner over large deployments, critical system functions, such as admission control, are provided through a centralized point of decision that is contacted whenever a decision cannot be taken locally. For the sake of simplicity, we assume hereafter that this functionality is provided by a single gateway called *Designated GW*.

To reflect this design the QoSBroker and the QoSDispatcher are divided into two different sub-component with different scope: *local* and *global*, respectively. In the reservation phase we rely on a *QoSGlobalBroker* and on a set of *QoSLocalBrokers*. The former is one for TaaS instance, while one instance of the latter is deployed in every gateway. The same approach is adopted for the QoSDispatcher (*QoSGlobalDispatcher* and a set of *QoSLocalDispatchers*). The local components manage, in term of QoS, the thing services provided by things that are directly connected to the gateway where the component is deployed. The global components, instead, have a global view of all the thing services available in the TaaS instance and are involved only when a global view is required. The overall architecture is shown in Figure 5. The choice of which QoS module is in charge of managing the global view among the local cloud is out of the scope of this work, however we can assume, without lack of generality, that a distributed election procedure is adopted between gateways. Below we provide a detailed description of the two main procedures that are involved to provide QoS support to applications.

5.1 Reservation

The reservation procedure is involved when an application wants to negotiate a set of thing services each one with certain QoS requirements. The applications can access thing services from any gateway that is part of the TaaS instance; however, from the TaaS point of view, one and only one of the gateways is the application's gateway – GW_1 in sequence diagrams. First of all, the application sends an agreement request to its gateway. The agreement request encapsulates the TSID required by the application. The request is handled by the QoSLocalBroker that forwards it to the QoSGlobalBroker. The QoSGlobalBroker validates the agreement request and, if it is feasible,

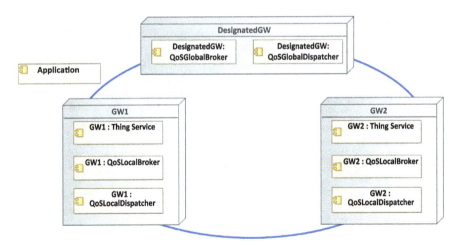

Figure 5 Deploy diagram

reserves resources for the application, performing so the admission control functionality. Then the QoSGlobalBroker replies with the AEPR, which is also stored by the QoSLocalBroker of the GW_1. The AEPR is forwarded back to the application in order to enable the application to invoke the negotiated thing services. For a detailed sequence diagram see Figure 6.

It is worth to mention that, in this phase, we do not perform the mapping between thing services and things. In fact, the time between the negotiation and the invocation phase is unknown, thus if we perform the selection in this phase this might result in a suboptimal allocation or incur in unpredictable errors due to the highly dynamic environment. However, the QoSGlobalBroker has the view of all the committed agreements, so it checks if the new agreement can be satisfied without violating previously accepted agreements. In other words, the QoSGlobalBroker checks if there is at least one allocation schema that can be adopted in order to fulfil the requirements of every agreement plus the new one. If the answer is positive, the new agreement is accepted; otherwise, the agreement is rejected and the application is notified. The application can obviously start a new negotiation phase with less stringent QoS requirements.

5.2 Allocation

The allocation procedure is performed every time an application invokes thing services. Applications can invoke thing services several times, thus the selected things can change between two different invocations. However,

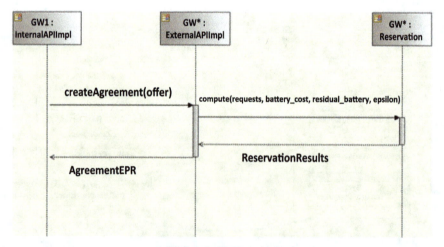

Figure 6 Reservation

there is a drawback in this approach, which is the overhead associated to each invocation. In order to overcome this limitation, we can perform the allocation phase only every k invocations reducing the computational overhead. Thus, there is a trade-off, modeled by the k parameter, between the optimal selection of things and the overhead associate to each allocation. To the sake of simplicity in the rest of the paper we assume $k = 1$, however the same considerations can be written for larger values of k.

Due the distributed nature of the TaaS, we can have two different types of allocation scenarios: *local allocation* and *global allocation*. The dispatcher uses the list of equivalent thing in order to discriminate between the two different procedures.

The local allocation is adopted when all the thing services requested are provided only by things directly attached to the same gateways of the application. In this case, shown in Figure 7, the overall process starts and ends in the local gateway without any external interaction. For the sake of explanation, we consider the case in which an application asks only for one thing service. First of all, the application invokes the thing service (TSID) with also the AEPR previously retrieved. These data are forwarded to the QoSLocalDispatcher, which validates the AEPR with the help of the QoSLocalBroker. If the QoSLocalDispatcher is authorized, it resolves the requested TSID to a TID that can be used to provide the thing service by exploiting the equivalent things list. Finally, the selected thing is involved and the results are sent back to the application.

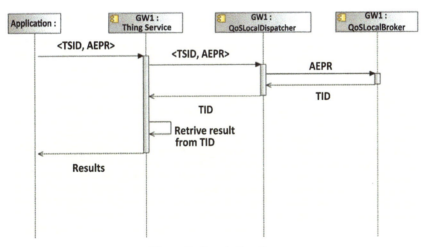

Figure 7 Local allocation

The global allocation is performed when the list of equivalent things contains at least one thing not attached to the application's gateway. In this case, the QoSLocalDispatcher must delegate the allocation procedure to the QoSGlobalDispatcher. The global allocation is split in two more cases: one that takes place when the things involved are attached to multiple gateways, and another one that takes place when all things, in the list of equivalent things, are attached to only one remote gateway.

The first case is explained in Figure 8. After the authorization interaction, the QoSLocalDispatcher forwards the TSID to the QoSGlobalDispatcher that replies back with the selected TID. Finally, the Thing Services modules interact and the results are sent back to the application.

The second case, instead, is a mix of the local and the global allocation, because if the application is attached to the remote gateway the overall process will be accomplished with the local allocation procedure. However, because the gateway involved is not the application's gateway a central point of coordination is needed. The overall process is shown in Figure 9 and is similar to the first global allocation procedure, however when the QoSGlobalDispatcher receives the allocation request it must forward such request to the QoSLocalDispatcher of the remote gateway. The remote instance of the QoSLocalDispatcher performs the allocation process following the local allocation procedure and then replies back with the selected TID. The QoSGlobalDispatcher sends the response to the origin QoSLocalDispatcher that forwards the TID to the Thing Service module. After the interaction between the Thing Service modules involved the results are sent back to the application.

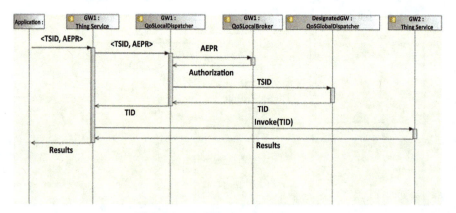

Figure 8   Global allocation, first case

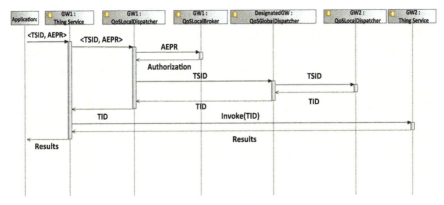

Figure 9   Global allocation, second case

6 Conclusions

In this paper a framework to support QoS specifically designed for IoT systems is presented. The proposed solution aims at introducing in large heterogeneous IoT systems functionalities for QoS negotiation and support. Specific attention to scalability is given in the design at any rate, while support for M2M applications characterized by heterogeneous QoS requirements is supported through Service Level negotiation. Eventually, the proposed platform establishes a fertile soil for the definition of algorithms and schedulers that can guarantee efficient resource allocation, which is left for future work.

7 Acknowledgement

This work has been carried out within the activities of the project "Building the Environment for the Things as a Service (BETaaS)", which is co-funded by the European Commission under the Seventh Framework Programme (grant no. 317674).

References

[1] P. Spiess, S. Karnouskos, D. Guinard, D. Savio, O. Baecker, L. Moreira Sá de Souza, and V. Trifa. "SOA-Based Integration of the Internet of Things in Enterprise Services," in Proceedings of the 2009 IEEE International Conference on Web Services (ICWS '09). 2009.

[2] "Machine-to-Machine communications (M2M); M2M service requirements," ETSI TS 102 689 V1.1.2.

[3] R. Q. Hu, Q. Yi, C. Hsiao-Hwa, and A. Jamalipour, "Recent progress in machine-to-machine communications," IEEE Communications Magazine. 2011.

[4] P. Morreale, F. Qi, P. Croft, R. Suleski, and B. Sinnicke, F. Kendall. "Real-Time Environmental Monitoring and Notification for Public Safety," IEEE Multimedia, 2010.

[5] C. Bennett and S. B. Wicker. "Decreased time delay and security enhancement recommendations for AMI smart meter networks," Innovative Smart Grid Technologies (ISGT). 2010.

[6] LOLA European project D3.6. "QoS metrics for M2M and Online Gaming".

[7] R. Liu, W. Wu, H. Zhu, and D. Yang, "M2M-Oriented QoS Categorization in Cellular Network," in 7th International Conference on Wireless Communications, Networking and Mobile Computing (WiCOM). 2011.

[8] M.-A. Nef, L. Perlepes, S. Karagiorgou, G. I. Stamoulis, and P. K. Kikiras. "Enabling QoS in the Internet of Things," in Proceedings of the Fifth International Conference on Communication Theory, Reliability, and Quality of Service. 2012.

[9] Z. Ming and M. Yan. "Modeling and Computational Method for QoS in IOT," in Proceedings of the 2012 IEEE 3rd International Conference on Software Engineering and Service Science (ICSESS). 2012.

[10] R. Duan, X. Chen, and T. Xing. "A QoS Architecture for IOT," in Proceedings of the 2011 International Conference on Internet of Things (iThings/CPSCom) and 4th International Conference on Cyber, Physical and Social Computing. 2011.

[11] "IEEE Standard Communication Delivery Time Performance Requirements for Electric Power Substation Automation," IEEE Std 1646–2004, 2005.

[12] O. Waeldrich, D. Battré, F. Brazier, et al.. "Web Services Agreement Negotiation Specification (WS-Agreement Negotiation)".

[13] O. Waeldrich. "WS-Agreement for JAVA (WSAG4J)," project website: http://packcse0.scai.fraunhofer.de/wsag4j

[14] E. Mingozzi, G. Tanganelli, C. Vallati, and V. Di Gregorio. "An open framework for accessing Things as a service," in the 16th International Symposium onWireless Personal Multimedia Communications (WPMC). 2013.

[15] V. Rajendran, K. Obraczka, and J. J. Garcia-Luna-Aceves. "Energy-efficient collision-free medium access control for wireless sensor networks," in Proceedings of the 1st international conference on Embedded networked sensor systems (SenSys '03). 2003.

[16] T. van Damand K. Langendoen. "An adaptive energy-efficient MAC protocol for wireless sensor networks," in Proceedings of the 1st international conference on Embedded networked sensor systems (SenSys '03). 2003.

[17] L. Yuan, W. Cheng, and X. Du. "An energy-efficient real-time routing protocol for sensor networks," Computer Communications. 2007.

[18] K. Akkaya andM. Younis. "An energy-aware QoS routing protocol for wireless sensor networks," in Proceedings of the 23rd IEEE International Conference on Distributed Computing Systems Workshops. 2003.

[19] T. Cucinotta, A. Mancina, G. F. Anastasi, G. Lipari, L. Mangeruca, R. Checcozzo, and F. Rusina, "A Real-Time Service-Oriented Architecture for Industrial Automation," IEEE Transactions on Industrial Informatics. 2009.

[20] D. Menascè, H. Ruan, and H. Gomaa, "QoS Management in Service-oriented Architectures," Performance Evaluation. 2007.

[21] F. C. Delicato, P. F. Pires, L. Pirmez, and L. F. Rust da Costa Carmo, "A Flexible Web Service Based Architecture for Wireless Sensor Networks," in Proceedings of the 23rd International Conference on Distributed Computing Systems (ICDCSW '03). 2003.

[22] G. F. Anastasi, E. Bini, A. Romano, and G. Lipari, "A service-oriented architecture for QoS configuration and management of Wireless Sensor Networks," IEEE Conference on Emerging Technologies and Factory Automation (ETFA). 2010.

[23] N. Shankaran, J. S. Kinnebrew, X. D. Koutsoukas, C. Lu, D. C. Schmidt, and G. Biswas. "An Integrated Planning and Adaptive Resource Management Architecture for Distributed Real-Time Embedded Systems," IEEE Transanction on Computers.2009.

[24] Y. Zhang, C. D. Gill,and C. Lu. "Configurable Middleware for Distributed Real-Time Systems with Aperiodic and Periodic Tasks," IEEE Transactions on Parallel and Distributed Systems. 2010.

Biographies

Enzo Mingozzi is an Associate Professor at the Department of Information Engineering of the University of Pisa, Italy, where he also received his PhD in Computer Systems Engineering. His research interests span several areas, including the design and performance evaluation of multiple access protocols for wireless (multi-hop/mesh) networks, quality of service provisioning in IP networks, and networking and services for the IoT. He has been involved in several national (FIRB, PRIN, PON) and international (Eurescom, EU) projects, as well as research projects supported by private industries (Telecom Italia Lab, Nokia, and Nokia Siemens Networks). He is co-author of 80+ peer-reviewed papers in international journal and conference proceedings, 4 book chapters and 10 patents. He currently serves on the Editorial Board of the Internet of Things Journal (IEEE), Computer Networks (Elsevier), and Computer Communications (Elsevier).

Giacomo Tanganelli is a PhD student at the Department of Information Engineering of the University of Pisa. He received a Master's Degree in 2012 from the University of Pisa. His main research interests include Internet of Things, Quality of Service, M2M communication, Service Oriented Architecture. He is currently involved in the FP7 European project BETaaS.

Carlo Vallati is a Postdoctoral Researcher at the Department of Information Engineering of the University of Pisa. He received a Master's Degree (magna cum laude) and a PhD in Computer Systems Engineering in 2008 and 2012, respectively, from the University of Pisa. In 2010, he visited the Computer Science department of the University of California at Davis. His main research interests include next generation broadband networks, wireless mesh networks, sensor networks, M2M communication. He is currently involved in the project BETaaS, Building the Environment for the Things as a Service, funded by the European Union under the 7th Framework Program. He has been involved in the past in research projects supported by private industries (Nokia Siemens Networks and Fluidmesh). He has served as a member of the organization committee for the international conference VALUETOOLS 2009 and WOWMOM 2011 and as member of the technical program committee for VEHICULAR 2012-13-14, Hotmesh 2013, ICACCI 2014 and VisioNet 2014. He has been the co-chair of IoT-SoS 2013 workshop co-organized with WoWMoM 2013.

Mediated Security Pairing for the Internet of Things

Jani Suomalainen

VTT Technical Research Centre of Finland

Received 28 May 2014; Accepted 28 June 2014
Publication 31 August 2014

Abstract

The Internet of Things (IoT) – global connectivity between all kinds of embedded devices and servers – is opening new opportunities for everyday applications. Essential enablers for the IoT are the secure and authenticated connections between things and servers. However, existing solutions for setting up thing-to-server authentication, based e.g. on passwords, trusted certification authorities, or physical connection, are not feasible when servers are far away and things do not have interfaces for inputting passwords or secrets keys. This paper analyses challenges and approaches for security pairing these interface restricted globally distributed things. We explore how mediating devices, such as smartphones, can be used to establish security connections. Particularly, we contribute by analysing how user-friendly security establishment approaches – out-of-band and unauthenticated location-based pairing – can be applied in situations where counterparty is far away or has incompatible interfaces.

Keywords: Internet of Things, embedded device, security, authentication, pairing, mediated protocol, smartphone.

1 Introduction

The Internet of Things (IoT) promises new opportunities for everyday living by allowing all kinds of devices to be connected with other devices and services

Journal of Communication, Navigation, Sensing and Services, Vol. 1, 129–150.
doi: 10.13052/jconasense2246-2120.122

anywhere in the world. The IoT is a promising communication approach for cyber-physical systems (CPS) where it enables various computational elements to cooperate and control physical entities. For example, one application for the IoT – a smart grid – promises energy savings by collecting information from sensors and actuators in cities and districts and then using this information to intelligently control energy production, transportation, and consumption. However, embedded IoT devices, with network connectivity but without displays or keyboards, are a challenge for security. How can the end-user securely connect a washing machine or lightning system to a previously unknown cloud service or remote controller when there are no means to input passwords?

Security pairing is a process that establishes an authenticated security relationship (often a shared secret key) for two devices without a prior relationship. This security relationship is then later used to protect the confidentiality, integrity and authenticity of communication. Reliable pairing mechanisms are needed to assure that devices communicate securely with those counterparties the user intended and not with man-in-the-middle attackers. To pair embedded devices easily, various mechanisms that are based on user's gestures and out-of-band (OOB) channels have been proposed and standardized [1, 2]. For instance, to pair two nearby sensors, the user can physically connect them (to touch one device with another or to connect them with a cable). Another easy solution for pairing devices is to utilize location information i.e. to pair devices in physical proximity. Unfortunately, these OOB and location-based security pairing solutions are unfeasible in the IoT when paired counterparties are far away. Also, even when devices are nearby, they need to have compatible interfaces. A device with NFC interface can be paired with another NFC device but cannot be paired with device with USB interface. A camera can be paired with a display but not with another camera.

In this paper, we explore how the end-user can – with a help of a mediating smartphone as illustrated in Figure 1 – create security associations with user interaction restricted IoT devices. In Section 2, we analyse requirements that IoT causes for authentication and recap existing solutions for authentication and pairing. We propose and analyse security pairing protocols for the IoT. We will focus on two particular challenge and case:

1. In Section 3, we will analyse and propose mediated protocols for solving interoperability challenges that exists when pairing devices with different (physical) OOB interfaces. Novel contributions include the analysis on the effects of OOB channels' directionality and type for pairing protocols.

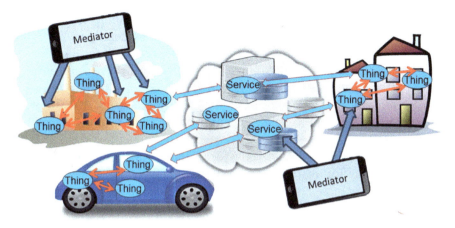

Figure 1 An overview of mediated pairing approach where smartphones are used to assist and control authentication between things and between services and things

2. In Section 4, we will propose a novel model for using context (location and time) information in the IoT to pair distant devices with a help of a mediating smartphone.

We then discuss feasibility and security assumptions of these proposals in Section 5 and summarize the contributions in Section 6.

This article is an extension to our conference paper [3], which was presented at Global Wireless Summit 2014. We have iterated and simplified the proposed protocols and provide real-world examples and further considerations on assumptions, feasibility, and interoperability of the proposed protocols. We have also extended our survey on the security and authentication challenges and solutions.

2 Authentication in the IoT

This section surveys existing solutions for authenticating devices in Internet and in short-range wireless networks and considers their applicability in the IoT.

2.1 Authentication in the IoT

Authentication is a process of ensuring the identity of communication counterparty. In the Internet, servers are typically authenticated using certificates, which are issued by trusted parties. For identifying devices in the IoT, the certification system does not scale. [4, 5] The certification process offered by third-party authoritiesis expensive and vulnerable as demonstrated by recent security incidents [6, 7]. Certificates may be infeasible for resource restricted devices as they require computationally expensive asymmetric crypto algorithms, sufficient bandwidth, and interfaces for installing them. Further, certificates give an ability to verify identity of servers that is not helpful if there is no mean to input domain name or address of the remote target.

Password-based authentication is a dominant solution for the user-authentication identification in Internet. An essential challenge in password based authentication in the IoT is how to register passwords (to remote party or, in case of federated single-sign-on solutions, to identity provider's servers) and how to deliver passwords or keys for devices.

One solution, used often e.g. with WiFi access points, is to preinstall a secret key to a device at the factory and then deploy the secret with the device e.g. in a printed sticker. The end-user is then required to provide the key for the remote counterparty using e.g. a PC or smartphone. Such solutions cause additional manufacturing cost due to setting unique keys and stickers to devices. Also, solutions are vulnerable for physical attackers, who are able to see visible keys, and for untrusted remote parties, as device may be connected to many remote targets but can only have limited amount of fixed keys printed.

Smartphones provide an alternative approach for registering IoT devices as they provide a secure channel to register and acquire (e.g. one-time) passwords, which are then inputted to an embedded device. However, these solutions are challenging when the devices do not provide sufficient interfaces for the end-user to type the password or copy a secret to device's memory.

2.2 Pairing in Short-range Wireless Networks

Secret key establishment between devices, which are personal and physically close, can be based on various mechanisms. Pairing standards for wireless short-range communication, including Wi-Fi Protected Setup, Bluetooth Secure Simple Pairing and Wireless Universal Serial Bus (USB) association models, have introduced the different mechanisms to complement the use of passwords or personal identification numbers (PINs).

- Short-string compare or input models are based on more recent security protocols [8–14], which allow the user to pair devices by comparing two displayed short strings or by entering the string on one device. Protocols provide strong secrecy against passive attacks and assume the user to be capable to prevent active guessing attacks that require physical access to devices. For IoT scenarios they are less suitable when the user cannot see values displayed by remote device or to control amount of guesses.

- Out-of-band (OOB) channel based pairing mechanisms use trusted channel to transfer critical parameters, which are later utilized in in-band channel to finalize the pairing. The OOB channels are typically not suitable for connecting a remote server with a local thing. Things may support different OOB interfaces, however different types of interfaces interoperable. Hence, OOB channels are not directly suitable as a generic pairing approach for the IoT.

2.3 Out-of-band Pairing Protocols

The seminal idea for OOB based pairing was presented by Stajano in the paper titled 'the resurrecting duckling' [15]. Afterwards various OOB channels have been proposed including Near Field Communication (NFC), Universal Serial Bus (USB) memories, cables, audio, and visible light (e.g. taking a picture from a displayed barcode or sensing LED pattern) [1]. Examples of devises with OOB interface are listed in Table 1. The table classifies devices according to the interfaces ability to transmit (out-bound), receive (in-bound), or both transmit and receive (bidirectional) pairing information.

Different protocols can be used for mutual pairing. The type of protocol depends on the security properties of OOB channel. Some channels are authentic, which means that an attacker cannot modify or prevent information transmissions. Some channels are confidential meaning that the information is secret and inaccessible for the eavesdropper. The level security (i.e. the work amount that must be done for a successful attack) depends on the actual implementation and assumptions on the used environment. Some examples of OOB channels in these categories are listed in Table 2. Typically the confidential channels are also authentic.

When the channel is authentic and confidential, a shared key may be directly passed through the OOB channel. The protocol is illustrated in

Table 1 Classifying out-of-band (OOB) channels with different directionalities

OOB Directionality	Examples of Channel Components
Out-bound	Display, blinking led, speaker, writing to USB memory
In-bound	Camera, keyboard, reading from USB memory, voice recorder
Bidirectional	NFC, USB cable

134   *J. Suomalainen*

Table 2   Classifying out-of-band (OOB) channels with through security properties

OOB Channel Security Properties	Examples of OOB Channels
Authentic	• Visible light(e.g. display to camera)
	• Sound (assuming that the user can detect attacks)
	• NFC (assuming that attackers are distant or when protecting integrity of NFC e.g. with iCodes[16])
	• Physical cables
	• USB memories
Confidential	• NFC (when assuming that there are no attackers near)
	• Physical cables
	• USB memories

Figure 2. In the following illustrations of protocol messaging, we use symbol '≡>' to denote transmission via authentic and confidential channel OOB channel, '=>' to transmission via authentic but non-confidential channel OOB channel, '->' transmission via in-band channel, and 'Key_{AB}' as symmetric key between Alice and Bob.

When the channel is two-directional and authentic (but not necessarily confidential), the devices may exchange their public keys (*PK*) and later on use in-band channel to agree a shared secret by exchanging Diffie-Hellman key agreement parameters (*DH*). The protocol, presented below, authenticates Diffie-Hellman protocol by signing parameters with the private keys. The protocol is illustrated in Figure 3.

When the channel is authentic but only one directional, the OOB channel provides only one-directional authentication. Mutual authentication requires an additional verification with the help of the end-user and support for device with outbound interface. SaxenA et al. [17] demonstrated how the protocol can be achieved with a single one-directional oob message, which is used to verify the outcome of the Short Authentication String (SAS) protocol [10, 11, 13, 14]. In the SAS protocol, each device computes a short checksum from the messages exchanged during the key agreement protocol. If the two checksums are the same, the exchange is authenticated. A man-in-the-middle attacker

---

1. Alice generates a secret key Key_{AB} and delivers it to Bob through a (one or two-directional) confidential and authentic OOB channel

   Alice≡>Bob: Key$_{AB}$

Figure 2   Pairing with a confidential out-of-band channel

---

1. Alice and Bob exchange public keys PK_A and PK_B through authentic (non-confidential) OOB channel

    Alice=>Bob: PK_A

    Bob=>Alice: PK_B

2. Alice and Bob exchange Diffie-Hellman parameters though unsecure in-band channel.

    Alice->Bob: DH_A, sign (DH_A, $PrivK_A$)

    Bob->Alice: DH_B, sign (DH_B, $PrivK_B$)

Figure 3   Pairing with a two-directional authentic out-of-band channel

would have different public key and, hence, checksums would not match. One-directional authentic channel can now be used to deliver the checksum from Alice to Bob. Bob is there're able to verify that the pairing succeeded. Alice is vulnerable for spoofing attacks. Bob will detect if this attack occurs and is able Bob is able to abort the pairing. Bob may also indicate failure for the end-user. Therefore, Alice should wait for end-user confirmation before allowing pairing to occur. The model assumes that Bob is able to inform the end-user and that end-user is able to trigger Alice to accept pairing. The protocol is described in Figure 4.

---

1. Alice and Bob exchange public keys PK_A and PK_B, random strings R_A and R_B as well as a cryptographic hash of R_A

    Alice->Bob: PK_A, R_A, $h(R_A)$

    Bob->Alice: PK_B, R_B

2. Alice checks that received hash matches to hash calculated from the string. Then Alice and Bob compute checksums $f = f(PK_A, PK_B, R_A, R_B)$. Where f is a cryptographic hash function with output length suitable for OOB channel.

3. Alice sends checksum to Bob through authentic (non-confidential) OOB channel

    Alice=>Bob: f_A

4. Bob accepts pairing if f_A equals f_B and indicates success for the user (e.g. Bob displays a message).

5. Alice waits until the end-user confirms Bob acceptance (e.g. by clicking ok).

Figure 4   OOB verified SAS protocol for pairing over one-directional authentic out-of-band channel

Bandwidth of OOB channel affects also to the protocols. In low-bandwidth OOB channels, passing a complete public key may impair the user experience. In these cases it is enough that devices only transmit a commitment to the public key using OOB channel as proposed by Balfanz et al. [18]. The commitment is a hash of public key and can be sent also on bandwidth limited OOB channels; when the actual public key is send though in-band channels. The commitment is then used to verify the authenticity of the public key.

Requirements for OOB channels have been analyzed and surveyed e.g. by Asokan et al. [19] who considered bandwidth, directionality, integrity and confidentiality and potential pairing protocols. However, they did not consider cases where the OOB channel is mediated i.e. consists of two different OOB channels.

Mediating devices can be used to establish a key for devices that cannot be directly connected with an OOB channel. Touch mediated Association Protocol (TAP) [20] is a solution where the end-user touches two devices with a third one in order to pair them. Tapping is based on forwarding secrets through a short-range wireless channel, which is assumed to be confidential.

2.4 Location-based Authentication

The paradigm of context-aware security – allowing a device to perform particular actions or adjusting device's security level based on contextual information such as location or time – has been an active research area in the scope of short-range networks and ubiquitous computing.

Proximity information – proves that devices are physically close to each other at particular time – has been used in security pairing by several researchers. Ekberg [21] proposed a solution for a mobile user carrying mobile phones and resource restricted sensors. The solution assumes attackers inability to follow the user and allows two devices to pair if they see each other several times in different places. Varshavsky et al. [22] and Mathur et al. [23] utilized observation that radio signals' phase and amplitude correlate with distance. Their solutions use characteristics of radio signals to verify that devices are in close proximity to each other. Paired devices first measure reciprocal signal characteristics, which are identical only for pair-wisely communicating devices, and then accept pairing only if the measurements are identical.

Location information is also utilized in many short-distance wireless standards to gain some level of security. WiFi Protected Setup [24] specifies 'PushButton' pairing model and Bluetooth Secure Simple Pairing [25] specifies 'JustConnect' pairing model. In both models, the end-user first conditions devices temporarily for pairing e.g. by pressing a button in both devices. Then the conditioned devices at that location and that time make the pairing with each other (if no other simultaneous pairing attempts are seen). These models are unauthenticated. The device will be paired with the man-in-the-middle attacker who is at the same location as the user and who is able to intercept devices pairing communication.

3 Protocols for Mediated Out-of-band Channel based Pairing

This section presents protocols for pairing a thing with an OOB interface with a remote service or with another thing with different OOB interface. Our work extends previous OOB protocols presented in Section 2.3. We use a smartphone, which is typically equipped with several OOB interfaces (like NFC, camera, display, USB port) to pair a device (with any-directional OOB channel) and a remote service or device - to mediate exchange of pairing information. Our solution is usable to pair a device with another device that has opposite directional OOB channels. Hence, we solve the interoperability problems with existing proposals: two devices that can only transmit through OOB channel or two devices that can only receive through OOB channel can now be paired with the proposed protocols.

3.1 Mediated Protocols

A summary of proposed protocols for making a pairing between two things with incompatible OOB interfaces is provide in the following tables. The tables try to find the simplest protocols that can achieve the pairing. They illustrate the effects that OOB channels' characteristics and combinations have for protocols. Particularly, we focus on the following questions:

- How a pairing can be organized if an OOB interface can either outbound (send) or inbound (receive)?
- How the type of OOB channel - i.e. authentic or both authentic and confidential - affects to the pairing?

The cases where both devices have confidential and authentic channels are straightforward and illustrated in Table 3. The secret key can be generated either in one device or a mediator and then passed to another or both via the mediator. Only in the case where both devices have outbound interfaces, the

Table 3   Mediated pairing protocols for devices with confidential OOB interfaces[1]

Alice \ Bob	Channel Type	Confidential and Authentic	
Channel Type	Direction	Send	Receive
Confidential and Authentic	Send	1) $A{\equiv}{>}M$: K_{AM}	1) $A{\equiv}{>}M$: K_{AB}
		2) $B{\equiv}{>}M$: K_{AB}	2) $M{\equiv}{>}B$: K_{AB}
		3) $M{-}{>}A$: $F(K_{AB},K_{AM})$	
	Receive	1) $B{\equiv}{>}M$: K_{AB}	1) $M{\equiv}{>}A$: K_{AB}
		2) $M{\equiv}{>}A$: K_{AB}	2) $M{\equiv}{>}B$: K_{AB}

mediator must first establish a secret with one device and then use this secret to secure the delivery of the final key.

The cases where both devices have only authentic non-confidential channel require more communicating and computing. These cases are illustrated in Table 4.

In cases where one device is able to send and one to receive, the mediator is used as an OOB channel to transmit information form device to another. The protocol may be based on SAS protocol, which was described in Figure 4: Alice and Bob first negotiate a short string and then the mediator is used to pass a checksum, which verifies that both devices know the same string and consequently have received untampered public keys.

In cases where devices have the same directional non-confidential OOB channel, the mediator must first establish secret communication channels with Alice and Bob. These individual device-to-mediator channels are formed with own OOB-verified SAS pairing negotiations. The mediator first negotiates a confidential channel with Alice and then with Bob. Then the mediator generates a secret key, which it can transmit securely for Alice and Bob.

Solutions based on SAS protocol are vulnerable for spoofing attacks. They require that the end-user to verify that the pairing has succeeded. In the 'send-send' case, setup the mediator must have a display and Alice and Bob input capability (buttons) for providing verification. In the 'receive-receive' case, Alice and Bob must have displaying capability and the mediator input capability. In the 'send-receive' cases, the receiver must be able to display

---

[1]In each table, we use the following notations: 'A' and 'B' are devices Alice and Bob, 'M' is mediator, arrows are transmission via untrusted ('$->$'), authentic ('$=>$'), or confidential and authentic ('$\equiv>$') OOB channel, 'K_{XY}' is symmetric secret between X and Y, 'PK_Y' is a public key of Y, '$F(\text{data}, \text{key})$' is an encryption and authentication function, and SAS is the protocol described in Figure 6.

Table 4 Mediated pairing protocols for devices with authentic OOB interfaces

Alice \ Bob	Channel Type	Authentic	
Channel Type	Direction	Send	Receive
Authentic	Send	1) M<->A: SAS protocol	1) A<->B: SAS protocol
		2) A=>M: checksum	2) A=>M: checksum
		3) M indicates success	3) M=>B: checksum
		4) the user confirms A	4) B indicates success
		5) M<->B: SAS protocol	5) the user confirms A
		6) B=>M: checksum	
		7) M indicates success	
		8) the user confirms B	
		9) M->A: $F(K_{AB}, PK_A)$	
		10) M->B: $F(K_{AB}, PK_B)$	
	Receive	1) A<->B: SAS protocol	1) M<->A: SAS protocol
		2) B=>M: checksum	2) M=>A: checksum
		3) M=>A: checksum	3) A indicates success
		4) A indicates success	4) M<->B: SAS protocol
		5) the user confirms B	5) B=>B: checksum
			6) B indicates success
			7) the user confirms M
			8) M->A: $F(K_{AB}, PK_A)$
			9) M->B: $F(K_{AB}, PK_B)$

the notification and the sender must have capability (a button) for the user to verify that pairing occurred successfully.

Cases where one device has authentic and one device has confidential channel are mixtures of the previous cases. These cases are illustrated in Table 5. The mediator must establish first a confidential in-band channel with the device that provides only authentic OOB interface. This is achieved with OOB verified SAS protocol. Then the mediator forwards the secret key for devices using confidential out-of-band channel and established confidential in-band channel.

The direction of transmissions and requirements for devices user interaction capabilities depend on the case. In the 'send-send' and 'send-receive' cases, the mediator have a display for indicating that SAS checksum check

Table 5   Mediated pairing protocols for cases with authentic OOB and confidential OOB

Alice \ Bob	Channel Type	Confidential and Authentic	
Channel Type	Direction	Send	Receive
Authentic	Send	1) M<−>A: SAS protocol 2) A=>M: checksum 3) M indicates success 4) the user confirms A 5) B≡>M: K_{AB} 6) M−>B: $F(K_{AB}, PK_B)$	1) M<−>A: SAS protocol 2) A=>M: checksum 3) M indicates success 4) the user confirms A 5) M≡>B: K_{AB} 6) M−>B: $F(K_{AB}, PK_A)$
	Receive	1) M<−>A: SAS protocol 2) M=>A: checksum 3) A indicates success 4) the user confirms M 5) B≡>M: K_{BM} 6) M−>A: $F(K_{AB}, PK_A)$	1) M<−>A: SAS protocol 2) M=>A: checksum 3) A indicates success 4) the user confirms M 3) M≡>B: K_{AB} 4) M−>A: $F(K_{AB}, PK_A)$

succeeded and Alice (with authentic OOB channel) must be able to receive user confirmation. In the 'receive-receive' and 'receive-send' cases, the situation is vice versa.

3.2 IoT Examples

To illustrate the protocols, this subsection presents two typical IoT case examples in more detail. We first consider a case where a camera is connected to a remote photo sharing service. Then we consider a case where a sensor with NFC interface is paired with an IoT gateway (which may forward sensor data to remote servers) with keyboard interface.

The smartphone is assumed to have a confidential and authentic connection to remote servers. This connection can be based secured using e.g. HTTPS. The end-user has registered to the service beforehand and acquired necessary credentials enabling the server to authenticate the smartphone user. The remote service is authenticated using certificates for the smartphone.

In the first scenario, illustrated in Figure 5, we have a camera that can take pictures from smartphone's display. This display is assumed to visible any one. So the scenario has an authentic inbound OOB channel and a two-directional confidential channel. Hence, we could select either protocol from the bottom row of Table 5. In the Figure 5, we selected the left alternative

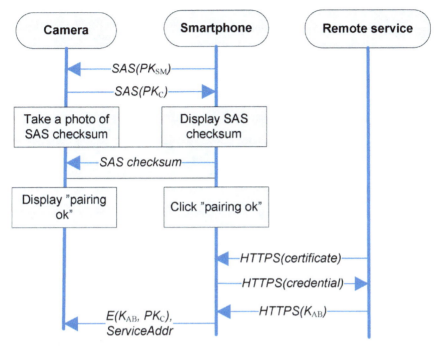

Figure 5 Pairing a camera and a remote service

where a symmetric key is generated in the remote server and then delivered to the sensor. The end-user first establishes a confidential and authentic connection with the sensor by executing the SAS protocol. The protocol is verified by taking a picture with a camera from smartphones display, which is showing the SAS checksum. As a result of SAS, the camera and smartphone know each other's private keys. In the next phase, the smartphone authenticates to the remote server and receives through the HTTPS secured channel the secret key. In the final phase, the smartphone protects the secret key with camera's public key and delivers the key to the camera.

In the second scenario, we have a sensor with NFC connection and a gateway with a keyboard. The NFC is assumed to be confidential. Gateway's keyboard is used to input string displayed by a smartphone. This visual channel is assumed to be authentic but not necessarily confidential. We select the protocol 'confidentialreceive' ->'authentic receive' from Table 5. The protocol for the scenario is presented in Figure 6. First the smartphone establishes a secure connection with the gateway by executing the SAS protocol. The pairing is verified by the end-user who types the SAS checksum displayed

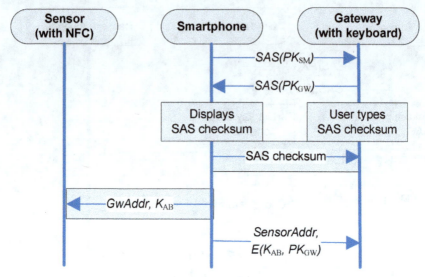

Figure 6   Pairing a sensor and an IoT gateway

by smartphone to the gateway. Then the smartphone generates random secret, which it delivers for the sensor and the gateway. The delivery to the gateway is protected with the gateway's public key that was learnt in the SAS negotiation. The delivery to the sensor occurs through an NFC interface. Additionally, also addressing information is delivered to the devices.

4 Location-aware Pairing with Distant Devices

This section presents an approach for pairing a local thing with a remote network server or thing with a help of a smartphone. The model is based on location and time and assistance from a trusted network service. Both the local thing and the smartphone are assumed to be aware of their own location. The model does not require any connectivity between smartphone and things. The approach is inspired by the unauthenticated location based pairing models (particularly WiFi Protected Setup [24] 'PushButton' and Bluetooth Secure Simple Pairing [25] 'JustConnect'), which were presented in Subsection 2.4. We extend applicability of these models into global IoT scenarios.

The location and time-aware protocol for pairing an embedded thing with a remote entity is illustrated in Figure 7. The thing is paired to the remote service when it is at the same location as a trusted smartphone and that

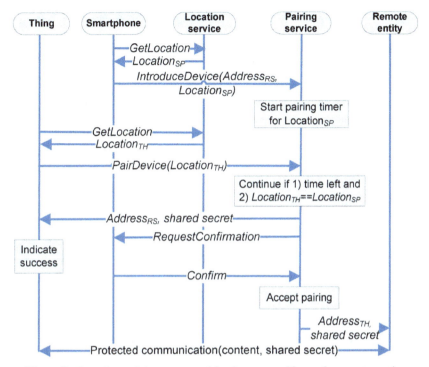

Figure 7   Location and time-aware pairing between a thing and a remote service

there are no other devices wishing to make a pairing at that same location at the same time. The solution is assisted with a smartphone and a pairing service (provided e.g. by device manufacturer or mobile network operator).

The protocol has the following phases:

1. To make a pairing, the user first conditions a smartphone and then the thing into a state where they want to make a pairing – e.g. by pressing a button or by connecting a power supply.
2. The smartphone (application) connects to the pairing service and sends an 'introduce service' request, which contains smartphone's location and address of the remote service (or e.g. another preregistered embedded device) to which the embedded device should be connected.
3. The pairing service starts a short, e.g. one minute, timer.
4. If a 'pair device' request from the same location is received before the timer runs out, the service replies with the address of cloud service and a secret key. If multiple pairing requests are received, the pairing is aborted.

5. After receiving the key, the device must indicate the success e.g. with a blinking LED or a sound beep.
6. The end-user, observing the indicator, sends a confirmation with a smartphone.
7. The pairing service, who now knows that a correct thing has received the secret key, forwards the key also to the cloud service along with the address information.

Devices must resolve their location. The available positioning mechanism affects to costs and security level that the approach provides. The positioning approach can be e.g.:

- Satellite positioning provides precise information and, thus, reduces risks that multiple devices are paired at the same location at the same time. But the costs of GPS receivers may be unfeasible for most IoT devices.
- Mobile network based location services are available without additional hardware for devices with mobile connectivity.
- Fingerprinting techniques (see e.g. [26]) can be used to identify nearby base stations or access points. Fingerprinting can be done at the physical communication layer by extracting unique signatures from wireless signals by looking amplitude, frequency, delays, and phases.

The location-aware pairing has some security assumptions. There must be a trusted pairing service. Certificates required for authenticating this service must be preinstalled to the devices or downloaded with a smartphone application. The embedded device must be equipped with some kind on indication mechanisms (e.g. LED, small display, or beep) and the user must be educated to check it. If this check is not made, an attacker may prevent legitimate thing from connecting the service and masquerade itself as the thing that user introduced.

5 Feasibility and Security Considerations

A major motivation for the proposed mediated pairing models comes from the interoperability. Use cases for pairing include e.g. connections between sensor or actuator and a cloud service; a television and cheap speakers (with NFC interface); two displays; as well as a device with USB port and a camera without any interface. These things are difficult to pair securely (without the risk of active man-in-the-middle attack), as the OOB interfaces are incompatible.

The incompatibility of OOB channel is not the only reason to use mediators. Two devices may have compatible OOB interfaces but it still may be more usable to use mediator instead of direct OOB connection. For instance, consider a case where NFC and Bluetooth enabled television and NFC and Bluetooth enabled air-conditioning system, which starts to blow when a storm is displayed in television. These devices may be located so that they cannot be connected with NFC. Due to their weight, they cannot be moved and paired. In these cases, a mobile phone acting as a mediator provides an easy alternative for making the pairing. The mediator-based pairing enables manufacturing of secure devices with less or cheaper hardware.

Backward compatibility causes some practical challenges for scenarios when mediator is used with standards that already provide some support for OOB pairing. For instance, in Bluetooth Secure Simple Pairing, paired devices exchange information on their I/O capabilities. The implementations are not aware of mediator and therefore the OOB interfaces will seem to be incompatible and the pairing process will stop. This problem may be addressed in two ways: at least another device is required to be compatible with our protocol and, hence, able to advertise that it has a compatible interface or the mediator must be able to intercept and modify capability negotiation messages, which are send through unencrypted in-band channel, so that the mediator becomes transparent.

Security levels of the protocols depend of different factors. Each protocol assumes a trustworthy smartphone. Further, each protocol relies on trustworthy connections. Confidential or authenticity levels of different OOB channels may vary: NFC is vulnerable for attackers in physical proximity; USB memories may get lost and forgotten; and USB cables are vulnerable only against advanced attackers capable of tampering physical connectors.

The location based pairing schemes depend on the reliability and security provided by the positioning mechanism. With some efforts it is possible to provide spoofed locations for devices. For instance, location information in mobile networks comes through authentic channels and the mobile network can authenticate each terminal. However, the mobile network cannot be completely sure that the signal from authenticated terminal is coming from the spot that the terminal locates. An attacker could 'tunnel' of signals to a distant base station with radio relays. A device, which is communicating with a (authentic) remote base station, will seem to be in different location than the intended pair. Consequently, the attacker may create pairing with both devices and launch a man-in-the-middle attack.

6 Conclusions

Security is an essential requirement for the IoT. A big challenge is how to connect embedded user-interaction restricted devices to remote servers or devices cost-efficiently, easily, and securely by utilizing available interfaces. Existing research with pairing mechanisms on short range wireless networks has proposed easy solutions based on OOB channels and unauthenticated context-aware solutions (based on location and temporal conditioning of device by the user). In this paper, we analysed how these mechanisms can be applied in typical IoT cases where counterparty is far away or has incompatible OOB interface. We contributed by analysing how directionality and type of complex OOB channel affects to authentication possibilities and protocols. Implementations and trials based on the presented designs are left for the future work.

References

[1] A. Kumar, N. Saxena, G. Tsudik and E. Uzun. 'Caveat eptor: A comparative study of secure device pairing methods'. Proceedings of the IEEE International Conference on Pervasive Computing and Communications. 2009. pp. 1–10.

[2] J. Suomalainen, J. Valkonen and N. Asokan. 'Standards for Security Associations in Personal Networks: A Comparative Analysis'. International Journal of Security and Networks 2009, Vol. 4, No. 1/2, pp. 87–100.

[3] J. Suomalainen. 'Smartphone Assisted Security Pairings for the Internet of Things'. Proceedings of the Second International Conference on Communications, Connectivity, Convergence, Content and Cooperation (IC5 2014). Aalborg, Denmark, 2014.

[4] R. Roman, P. Najera and J. Lopez. 'Securing the internet of things'. Computer 2011, Vol. 44, No. 9, pp. 51–58.

[5] T. Heer, O. Garcia-Morchon, R. Hummen, S. L. Keoh, S. S. Kumar and K. Wehrle. 'Security Challenges in the IP-based Internet of Things'. Wireless Personal Communications 2011, Vol. 61, No. 3, pp. 527–542.

[6] G. Keizer. 'Hackers may have stolen over 200 SSL certificates'. Computerworld. 2011. http://www.computerworld.com/s/article/9219663/Hackers_may_have_stolen_over_200_SSL_certificates.

[7] Mills, E. 'Comodo: Web attack broader than initially thought'. CNET, 2011. Available: http://news.cnet.com/8301-27080_3-20048831-245.html.

[8] J. Larsson. 'Higher layer key exchange techniques for Bluetooth security'. Open Group Conference. 2001.

[9] C. Gehrmann, C. Mitchell and K. Nyberg. 'Manual Authentication for Wireless Devices'. RSA CryptoBytes 2004, Spring, Vol. 7, No. 1, pp. 29–37.

[10] P. R. Zimmermann. 'PGPfone: Pretty Good Privacy Phone Owner's Manual, Version 1.0 beta 5, Appendix C'. 1996.

[11] S. Laur, N. Asokan and K. Nyberg. 'Efficient Mutual Data Authentication Using Manually Authenticated Strings. IACR Cryptology ePrint Archive'. 2005.

[12] S. Vaudenay. 'Secure Communications over Insecure Channels Based on Short Authenticated Strings'. Advances in Cryptology - CRYPTO 2005. Vol. 3621. 2005. Lecture Notes in Computer Science. pp. 309–326.

[13] S. Pasini and S. Vaudenay. 'SAS-based Authenticated Key Agreement'. Proceedings of The 9th International Workshop on Theory and Practice in Public Key Cryptography. Vol. 3958. 2006. Lecture Notes in Computer Science. pp. 395–409.

[14] M. Cagalj, S. Capkun and J. Hubaux. 'Key Agreement in Peer-To-Peer Wireless Networks'. Proceedings of the IEEE 2006, Vol. 94, No. 2, pp. 467–478.

[15] F. Stajano and R. Anderson. 'The Resurrecting Duckling: Security Issues for Ad-hoc Wireless Networks'. Proceedings of the 7th International Workshop on Security Protocols. 1999. Lecture Notes in Computer Science. pp. 172–194.

[16] M. Cagalj, J. Hubaux, S. Capkun, R. Rengaswamy, I. Tsigkogiannis and M. Srivastava. 'Integrity (I) Codes: Message Integrity Protection and Authentication Over Insecure Channels'. Proceedings of the 2006 IEEE Symposium on Security and Privacy. 2006. pp. 280–294.

[17] N. Saxena, J. Ekberg, K. Kostiainen and N. Asokan. 'Secure Device Pairing based on a Visual Channel (Short Paper)'. Proceedings of the 2006 IEEE Symposium on Security and Privacy. 2006. pp. 306–313.

[18] D. Balfanz, D. K. Smetters, P. Stewart and H. C. Wong. 'Talking to strangers: authentication in ad-hoc wireless networks'. Proceedings of the Network and Distributed System Security Symposium. 2002.

[19] N. Asokan & K. Nyberg. 'Security Associations for Wireless Devices'. Anonymous Security and Privacy in Mobile and Wireless Networking. Leicester, UK: Troubador Publishing Ltd, 2009.

[20] A. Lakshminarayanan. 'TAP - practical security protocols for wireless personal devices'. Proceedings of 15th IEEE International Symposium on Personal, Indoor and Mobile Radio Communications. 2004.

[21] J. Ekberg. 'Key establishment in constrained devices'. Seminar on Authentication and Key Establishment. Helsinki University of Technology. 2006. http://www.tcs.hut.fi/Studies/T-79.7001/2006AUT/seminar-papers/Ekberg-paper-final.pdf.

[22] A. Varshavsky, A. Scannell, A. LaMarca and E. d. Lara. 'Amigo: Proximity-based Authentication of Mobile Devices'. Proceedings of the Ninth International Conference on Ubiquitous Computing. Vol. 4717. 2007. Lecture Notes in Computer Science. pp. 253–270.

[23] S. Mathur, R. Miller, A. Varshavsky, W. Trappe and N. Mandayam. 'Proximate: proximity-based secure pairing using ambient wireless signals'. Proceedings of the 9th international conference on Mobile systems, applications, and services. 2011. pp. 211–224.

[24] Wi-Fi Alliance. 'Wi-Fi Protected Setup'. Web site. 2014. http://www.wi-fi.org/knowledge-center/faq/how-does-wi-fi-protectedsetup-work;.

[25] Bluetooth Special Interest Group. 'Bluetooth specification version 2.1'. 2007. www.bluetooth.com/NR/rdonlyres/F8E8276A-3898–4EC6-B7DA-E5535258B056/6545/Core_V21_EDR.zip.

[26] B. Danev, D. Zanettiand S. Capkun. 'On physical-layer identification of wireless devices'. ACM Computing Surveys (CSUR) 2012.

Biography

Jani Suomalainen is a senior scientist working in the cyber security team at VTT Technical Research Centre of Finland. He holds a Master of Science degree from Lappeenranta University of Technology and a Licentiate of

Science degree from Aalto University, Helsinki. He has worked in VTT since 2001 and participated to different communication platform and security related projects. His current research interests include authentication and authorization in ubiquitous and heterogeneous networks. He has authored around twenty academic articles and contributes actively to conferences and journals as a peer reviewer.

Adaptive Monitoring and Control Architectures for Power Distribution Grids over Heterogeneous ICT Networks

Rasmus L. Olsen[1], Christian Hägerling[3], Fabian M. Kurtz[3],
Florin Iov[2] and Christian Wietfeld[3]

[1]Department of Electronic Systems, Networking and Security,
Aalborg University, Denmark
[2]Department of Energy Technology, Aalborg University, Denmark
[3]Communication Networks Institute, TU Dortmund University,
Germany

Received 28 May 2014; Accepted 28 June 2014
Publication 31 August 2014

Abstract

The expected growth in distributed generation will significantly affect the operation and control of today's distribution grids. Being confronted with short time power variations of distributed generations, the assurance of a reliable service (grid stability, avoidance of energy losses) and the quality of the power may become costly. In this light, Smart Grids may provide an answer towards a more active and efficient electrical network. The EU project SmartC2Net aims to enable smart grid operations over imperfect, heterogeneous general purpose networks which poses a significant challenge to the reliability due to the stochastic behaviour found in such networks. Therefore, key concepts are presented in this paper targeting the support of proper smart grid control in these network environments. An overview on the required Information and Communication Technology (ICT) architecture and its functionality is provided and a description of one of several use cases, the External Generation Site is detailed along its evaluation approach.

Journal of Communication, Navigation, Sensing and Services, Vol. 1, 151–180.
doi: 10.13052/jconasense2246-2120.123

Keywords: Distribution Grids, Dispersed Generation, Smart Grids, Heterogeneous Communication Networks, AMR, Control, SmartC2Net.

1 Background and Motivation

Future electrical grids are characterized by a large diversity of renewable and distributed energy resources which, if not properly controlled, poses a challenge for the electrical grid. To ensure a reliable operation of the electrical grid while using heterogeneous networks, the FP7 project SmartC2Net [2, 11] employs a concept of two interacting control loops. As shown by Figure 1 the inner communication control loop deals with the intricacies of managing networks and their related protocols. It is thus in charge of handling the smart grid control messages which are issued by the outer energy control loop. The energy control algorithms of this outer loop work on input data collected from the grid's energy sensors and send their results back to the appropriate actors via the respective communication networks. The interaction of both loops, energy as well as ICT control, is a critical part of the project as it enables their coordinated operation to achieve the desired adaptivity and robustness of the Smart Grid even in the presence of faults and attacks.

The outer energy control loop performs monitoring and control tasks for the numerous distributed components of the grid. It therefore needs a means for communication enabling the reception of sensor data from field devices,

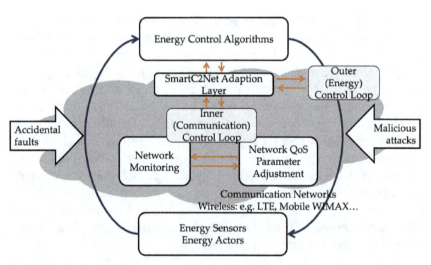

Figure 1   SmartC2Net concept of interacting energy and ICT control loops, [3]

as well as the transmission of control signals. Due to the different locations of grid devices, varying communication infrastructures are in place for delivering data with given protocols. As reliability is of paramount importance the overall system needs to maintain a high level of service quality in cases of poor network performance, partial or even complete failures and cyber attacks. In order to demonstrate and evaluate the developed ICT platform's capabilities to operate safely under challenging conditions, several real-world use cases are in focus of the SmartC2Net project. Communication infrastructure solutions and middleware have become of interest in supporting such smart grids as surveyed in e.g. [10]. Here SmartC2Net takes into account a unique and tight interaction between the two loops and a close link to evaluation in several realistic test bed setups described in this paper.

1.1 Background and Motivation for Smart Grid Control

Campaigns from Europe'as national governments promoting green energies have led to an increased penetration of distributed Photovoltaic (PV) systems in households, especially in Germany and Southern Europe. Installation of small household wind turbines, newly supported in Denmark for example, will further increase the amount of fluctuating power in distribution grids. All these dispersed generation units can introduce power quality issues as well as bottlenecks and congestion. For example, some of the Danish Distribution System Operators (DSOs) have encountered problems with voltage profiles in feeders with high penetration of household PV installations. By adding more fluctuating power from small wind turbines, the present problems, i.e. poor voltage and power quality, will worsen, eventually triggering the disconnection of loads or entire parts of the distribution grid. In some cases power quality issues might be solved by curtailment of power. However, this can result in energy systems devoid from clean and renewable energy sources.

Typically, DSOs monitor Medium Voltage (MV) feeders and maintain admissible voltage levels in these feeders by means of On-Load Tap-Changers (OLTC) installed in primary substations (High Voltage to Medium Voltage). Thus voltage on a substation's MV side is shifted upwards or downwards in small steps according to measurements at critical points. However, this control is not able to run smoothly and very often as required by the intermittent fluctuating renewable sources. Capacitor banks, also installed in primary substations, may contribute with an additional reactive power injection, hence boosting the voltage in MV feeders. However fast transients are introduced into the grid when switching these capacitor banks on and off. Moreover, there are limitations regarding the number and the frequency of this switching imposed by technical requirements. Notwithstanding, all these control capabilities were not designed to cope with fluctuating renewable energy. The situation is more critical in secondary substations (MV to LV) where all the small dispersed generation units at household level are actually being installed. The DSOs are currently not automatically controlling the voltage profiles in these Low Voltage (LV) grids. In some cases they might only have some limited information regarding the load of these substations.

Currently renewable energy systems like small wind turbines and solar PVs are capable of providing smooth control of reactive power. However, this capability is not used in a coordinated manner by DSOs. The SmartC2Net project strives to activate these capabilities and to provide DSOs with more and better control options for heterogeneous communication networks.

1.2 An Introduction to Heterogeneous Smart Grid ICT Architectures

A subset of SmartC2Net's Smart Grid enabling communication technologies, fitted into an architectural structure, is given by Figure 2. Starting from the left, the three device groups industrial, commercial and residential pool those technologies which are most commonly used in the respective context.

Though not limited to homes, in terms of scale these three groups here exist on the Home Area Network level (HAN). They connect via gateways (GW) to the upper layers of the architecture, using such technologies as ZigBee Smart Energy, Wi-Fi, HomePlug, KNX and others as indicated by the box to their right. Depending on the usage scenarios examples for gateways on this level are such diverse devices like an Energy Management GW (EMG, the central device of a Customer Energy Management System, i.e. CEMS), Home Automation (HA), Customer Premises Equipment (CPE, i.e. Internet modem/router) or a Metering GW (henceforth referred to as Automated Meter Reading Gateway - AMRGW). All of those gateways enable access to the next higher layer, in this instance the Neighborhood Area Network (NAN), for which DLMS, SML and PRIME are common technologies. More options can be found in Figure 2 in the box above HAN level. Options for wireless access networks in this domain involve the standards GPRS, WiMAX and LTE for cellular RF technologies and Wi-Fi as well as ZigBee in terms of RF-Mesh communication. Wire based solutions encompass the fiber optic GPON, copper-wire standards like (Euro-) DOCSIS and DSL, as well as different varieties of Power Line Communications (PLC), which are advantageous

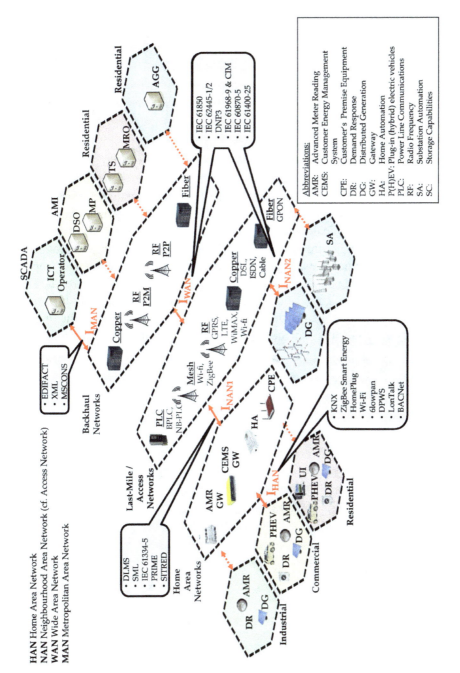

Figure 2   State-of-the-art technologies and standards for smart distribution grids [6]

in deployment as they operate via pre-existing power lines. All these options for access networks are also in use for connecting Substation Automation (SA) and Distributed Generation (DG) sites to the overall architecture and its higher layers.

Going further upwards in this hierarchy of communication networks one arrives at the WAN scale which is represented by Backhaul Networks. As their purpose is the interconnection of thousands of devices deployed over multiple, physically distinct sites there are high technological requirements concerning reliability, data rate, latency and more. Thus wired ICT is clearly predominant on this level, albeit wireless technologies, mostly comprised of high-grade technology with redundancy and dedicated spectrum in order to provide high reliability, can also be found. In communication technologies the achievable data rate typically decreases as range increases. This correlation is shown graphically by Figure 3, which includes a selection of widely employed access network technologies. Especially wireless technologies show a steep drop of the achievable data rate with growing distances, whereas wired ICT has a clear advantage. This is however associated with considerably higher deployment costs, which is the prime reason why fiber based infrastructures are not pervasive despite their obvious advantage in range and throughput capabilities.

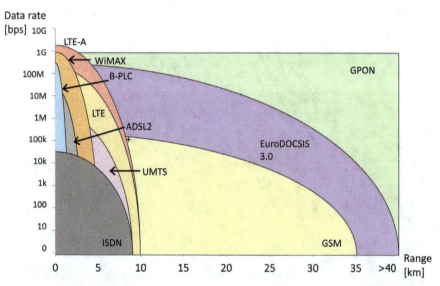

Figure 3   Range vs. Data rate of selected communication technologies [5]

2 Architecture and Use Cases

As it is impossible to foresee every potential error or to predict all conceivable avenues of attacks, resilience in the presence of malicious and accidental faults is a key requirement for Smart Grid ICT solutions. A means for coping with impaired services is therefore central to those functionalities which are essential for reliable grid operation. This includes the control, monitoring and communication aspects of Smart Grids. Moreover the deployment of communication networks dedicated to service only the energy grid is, among other obstacles, not financially feasible. It is thus unavoidable to make use of those pre-existing ICT infrastructures which are already deployed in the field. This leads to an overall heterogeneous infrastructure and thus to the necessity to account for the properties of such heterogeneous networks in the architecture's overall design. Special considerations need to be taken in accordance with the requirements of the individual use cases. In scope of the SmartC2Net project the following four use cases are considered [2]:

- **Medium Voltage Control (MVC):** focusing on medium voltage control even under cyber attacks.
- **External Generation Sites (EGS):** focusing on meeting demand and responses at medium and low voltage grids in different network conditions and technologies.
- **Electrical Vehicle Charging (EV):** focused around scheduling and planning the charging process of electrical vehicles in low voltage grids and at charging stations attached to the medium voltage grid.
- **Customer Energy Management Systems (CEMS)** and **Automated Meter Reading (AMR):** focusing on the integration of devices on customer sites.

Voltage Control in medium voltage grids with the aim of enabling and connecting Distributed Energy Resources (DERs), is the focus of the Medium Voltage Control use case. Architecturally it enfolds control facilities of the Transport and Distribution System Operators (TSO, DSO), DERs, primary substations and flexible loads. In light of the volatile power generation characteristics of DERs, with their associated dynamic power flows in distribution grids, a dependable communication of all the involved components is crucial to the system's reliable operation. This requirement is intensified as the loss or corruption of information, e.g. setpoints, might cause cascading effects thus putting the whole grid's stability at risk. Maximizing grid stability is achieved in the Medium Voltage Control use case via monitoring the distribution grid and using this information to one's advantage. From the collected data setpoints for DERs, flexible loads and medium to high voltage substations' equipment are calculated. This process is supported by forecasts for load, generation and weather conditions which help in increasing setpoint accuracy and thus grid stability by supplying valuable information that allows to mitigate negative effects from stochastic processes (i.e. generations and loads).

The project's second use case, External Generation Site, is concerned with decentralized energy storage and generation, aspects of Smart Grids that experience a steady rise in deployment and thus importance. Despite being primarily tasked with the control of low voltage grid entities, this use case also incorporates interfaces to mid and high voltage grid parts like secondary and primary substations. The reasoning behind this can be found in the benefits substation controllers might extract from an ability to communicate with external generation sites. Examples of such profitable information exchanges are the ability to pass the flexibility aggregated on the low voltage level on to MVC, as well as the optimization of energy costs, losses and low voltage profiles. Such functionalities depend on reliable information flows between the respective devices and are therefore enabled by resilient ICT infrastructures.

The circadian rhythm and concurrent working hours create high grid loads through synchronized charging patterns of Electrical Vehicles (EVs). Hence the EV Charging use case is introduced, where the vehicles connect to low voltage grids while additional charging stations exist on the medium voltage level. Here the flexibility in demand is harnessed to balance energy flows in the grid. As charging stations, EVs and the DSOs need to be capable of exchanging information in order to coordinated their tasks, a secure and reliable communication infrastructure becomes a key requirement.

AMR and CEMS are functionalities that are integral to Smart Distribution Grids. Out of this reason buildings in scope of this work are outfitted with Smart Metering devices which are capable of collecting consumption data, mainly concerning electric, gas, water and heating. This data can be used for tasks like customer feedback or transmission to the energy utilities where billing, accounting and balancing of the grid can be performed. AMRs enable this and also provide Demand Side Management (DSM). Direct DSM, load shifting to time slots with less grid load, more favorable pricing or improved utilization of local energy resources (i.e. DER), as well as added-value services are enabled through a CEMS. To achieve such a wide range of services and functionalities, the ICT architecture has to encompass all relevant household devices and connect them with appropriate access networks.

The unique sets of challenges associated with the four use cases outlined above are discussed in greater detailed in [2]. Figure 4 depicts all use cases considered in scope of SmartC2Net and their high level connections to several communication networks, which feature different characteristics in terms of

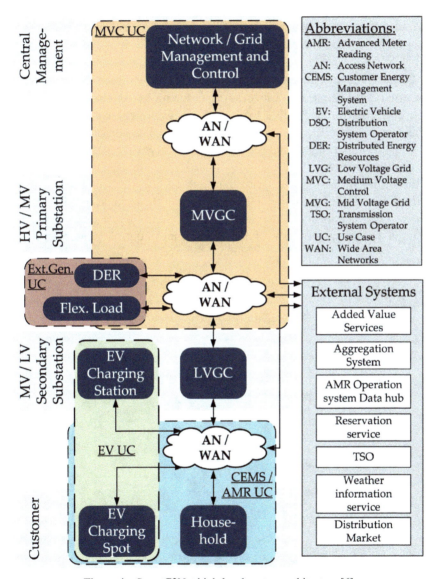

Figure 4 SmartC2Net high-level system architecture [6]

their properties including latency and data rate. These varying parameters affect the overall control of a Smart Grid as some specific, advanced algorithms might exceed the capabilities of the interfaces available.

2.1 Use Case Example: External Generation Site

Of the four different use cases in scope of the project, in this paper we discuss the external generation site use case in details.

2.1.1 Objectives of the use case

The external generation site use case is focusing on demonstrating the feasibility of controlling flexible loads and renewable energy resources in LV grids over an imperfect communication network. The flexibility provided by LV grids for upper hierarchical control levels is also investigated. The use case is captured in Figure 5.

In this setting the low voltage grid controller (LVGC) located in secondary substations should preferably be able to: 1) control the voltage profile along the low voltage feeders and 2) aggregate the LV flexibility that can be used as an input to the upper parts of the hierarchical control located in the

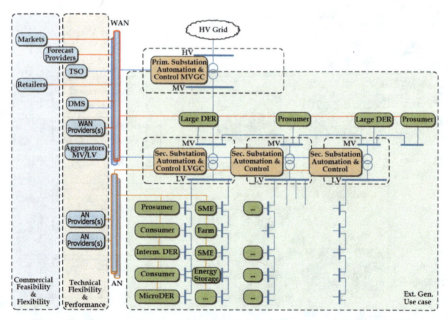

Figure 5 Communication network infrastructure imposed on the power grid [11]

primary substation (Medium Voltage Grid Controller, MVGC) and Distribution Management System (DMS). The medium voltage grid control should focus on: 1) control the voltage profile along the MV feeders, 2) minimize energy losses and 3) optimize energy costs. The grid operation should be resilient to faults and performance degradation in the public communication lines between the low voltage grid controller and the assets in the electrical grid with special focus on the low voltage side, hereby limiting the effect of changing network conditions on the electrical grid performance. This means that the use case also includes mechanisms for adapting the communication to events in the network that challenge the communication and the quality of the data exchanged between the controlled and controlling entities.

The complete grid operation should then be resilient to faults and performance degradation in the public communication networks. This resilience is achieved by (re-)configuration of communication network(s) and controllers involved in the process to accommodate for changing conditions and failures. This also means that this use case includes mechanisms for adapting the communication to degradation events in the network. As a part of this data quality affected by network performance is used as a key performance parameter to perform (re-)configurations of the network. In this context different sets of actors will interact to achieve two objectives:

- **Technical flexibility and performance:** Resilience of control towards faults and congestions in communication networks
- **Commercial feasibility and flexibility:** Aggregation of generation and demand (abstraction of models)

Figure 5 also shows the proposed communication network structure imposed on the power grid. The use case includes control of LV grid components, such as households, farms, PVs, local wind turbines, etc., as well as MV grid components such as larger refrigerator systems, wind farms and the like. A reliable communication is needed to ensure proper transport of measurement and control signals to assure proper control of the assets in the grid. Thus, the electrical grid includes both LV as well as MV grid. The communication network is split into three categories: 1) the access network (shown in orange) connects all actors in the low voltage grid; 2) the dedicated wide area network (shown in blue) which is a high performance dedicated network for grid control; 3) the wide area network (shown in red) connects the rest of the middle voltage actors.

A key challenge for data exchange over any general purpose networks is that these networks cannot be assumed to be designed specifically for the required task, and will be used by many other applications and services. Sharing the network means stochastic packet delays as well as packet losses from different handling of data packets throughout the network stack and network components. Stochastic delays and packet losses lead in turn to undesirable effects at the protocol level, and in turn affect the exchange of data information necessary for stable grid operation. Therefore, it is imperative to enable the smart grid systems to handle different situations in the communication network. This is done by monitoring the network's Quality of Service (QoS) and perform any (re-)configuration necessary possible. If there are no other alternatives then faults can be mitigated at the control level, e.g. by changing to a different and more conservative control strategy to gain grid stability at the cost of performance. To support the control system a range of concurrent network activities is therefore necessary: 1) monitoring of the communication network to keep track of what is going on in the communication network, 2) scalable management of data access mechanisms to cope with the potential number of sources and their geographical spread, 3) adaptation of QoS, protocols or network interfaces to allow continuous, undisturbed control. In relation to data access, data quality estimation is done to monitor the impact of the combined information dynamic and network performance. This is useful for internal reconfiguration and prioritisation of data management. An example related to context management can be seen in [12]. Finally, registration of communicating entities is needed for the system to be aware of the interacting entities. To scope all these aspect, the use case is divided into two orthogonal cases; control of assets and network adaptation.

2.1.2 Key performance indicators
To be able to measure the success of the overall combined control system, some indicators are denned, [2]:

- **Upper and lower limits in low voltage feeders:** The voltage profiles in the low voltage feeders shall not exceed +/– 10% from their rated value for a 10 min. average.
- **Dynamic upper and lower voltage limits in low voltage feeders:** The voltage profiles in the low voltage feeders shall not exceed +/– 3% of the rated value as a 3 second average value.

- **Loss reduction:** A loss reduction of 30% in the medium voltage grid compared with the base case (discussed in Section 2.1.3 without communication issues) shall be considered a success.

Similar, a set of key performance indicators for the communication will be evaluated. Later on, these can be set as requirements to smart grid solutions based on the control structure developed in SmartC2Net.

- **Packet losses:** How many percentages of packets send may be lost during transport.
- **Delay:** The delay limits for data to be send from signal source to the sinks (the controllers) and the return, for reference signals to be distributed to relevant controllable elements.
- **Throughput:** The amount of data inclusive overhead for managemen-tand adaptation being exchanged in the network over a defined time period.

2.1.3 Control of assets

Normal operation of the grid is defined by three sub cases that will be considered in situations where the effect of the network performance can be neglected:

- **Energy balance:** where the operation of MV grids is in scope. LV grids are considered aggregated and the Low Voltage Grid Controller (LVGC) is offering flexibility to the Medium Voltage Grid Controller (MVGC). Thus the MVCG is primarily controlling the assets such as large DER, prosumers and LV grid via the LVGC to keep the energy balance.
- **MV operation:** where the focus is to control the voltage profile as well as to optimize losses and energy costs on MV grids using active and reactive power capabilities offered by large DER, MV prosumers and the secondary substations on MV side.
- **LV control:** where the focus is to control the voltage profile on LV grids using reactive power capabilities offered by micro and intermediate DERs, flexible consumption and production at household or small and medium enterprises.

These three control cases serves as base line cases for cases where the network will start to potentially influence these operations. The three control cases described above therefore are orthogonal to network cases studies discussed in the following.

2.1.4 Network adaptation

The use cases for network adaptation focuses on (re-)configuration of the network(s) and protocol(s) to cope with network performance degradation. For this three fault/error sub-cases are defined:

- **Network Performance Changed:** This scenario deals with time varying performance in the network, and the adaptation of access methods to provide reliable data exchange between entities communicating.
- **Network Congestion:** This scenario deals with more severe network conditions, i.e. congestions in the network, and the adaptation of access methods to provide reliable data exchange between entities communicating.
- **Lost Network Connectivity:** This scenario addresses the case where devices loose connectivity at the network layer. The case assumes a certain notion of connectivity, e.g. as in TCP.

3 ICT Architecture and Functionalities

The overall architecture of SmartC2Net consists of several parts. The most obvious ones are the electrical grid and the ICT infrastructure, which each provide services without which the other could not be sustained. As the grid delivers power to ICT devices they enable the communication and thus control of individual grid parts. The communication side can be viewed from a logical and from a physical perspective. While the former highlights which devices are the sources and sinks of exchanged information the latter describes the underlying physical structure of the networks and thus shows which components are connected. The logical architecture represents a high level overview of SmartC2Net. Logical information flows and interfaces detailed in the project are mapped onto the underlying physical ICT architecture, i.e. the different ISO/OSI layers, of the project in [5].

Figure 6 gives a high level overview of the physical ICT architecture of SmartC2Net's use cases. While interfaces on the DSO level are mostly dedicated fiber lines, the other device groups, i.e. use cases, mainly communicate via public Wide Area Networks. This layout is due to the high costs associated with deploying dedicated high speed networks which is economical in select cases but not always for parts like external generation sites, especially if preexisting solutions can meet the use cases' requirements.

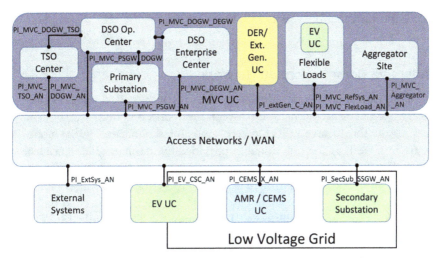

Figure 6   Communication network architecture overview [5]

The developed ICT platform that shall ensure reliable data exchange between the system components is closely related to the system architecture. As mentioned previously, the overall framework of the project comprises of several adaptation levels which will be elaborated in the following:

- Adaptive Monitoring
- Adaptive Grid Control
- Adaptive Communication Networks

3.1 Adaptive Monitoring

The part of the ICT which relates to monitoring is divided into four layers; 1) one focusing on monitoring of distributed variables which may be accesses via different, independent technologies and systems. 2) One dealing with processing and aggregation of data to ensure a scalable system solution. 3) One focused on publishing the information to the control system and other interested parties and finally 4) an API to ensure interaction of the different mechanisms envisioned related to fault management, grid control and network reconfiguration. One of the key features in this part will be the ability of the platform to provide reliability metrics of the information being accessed, allowing the adaptation mechanisms and control to take proper action instead of having to base their decisions on unreliable data. Internally this is also used to optimize resources spend on accessing the information versus reliability requirements, see [9] for more details.

3.2 Adaptive Communication Networks

The comprehensive network adaptation and reconfiguration approach of SmartC2Net enables an interaction between these different domains. For the network adaptation all seven ISO Open Systems Interconnection (OSI) model layers are involved, as the functionality provided by the inter related control loop includes all of them for maximum effectiveness of the QoS parameter adjustment as illustrated in Figure 7. The lower layers offer flexibility in terms of QoS, like frequency allocation in case of wireless technologies and native adaptations that are part of the respective standards. The middle ISO/OSI layers utilize Software Defined Networking for network reconfiguration and QoS network layer prioritisation. The upper layers of the SmartC2Net approach employ CIM-based information models and custom traffic classes. On the highest level, delay and loss characteristics are mapped into information reliability degradation metrics which, influenced by the access protocols, are used to (re-)configure and tune the protocols and QoS settings, see e.g. [12]. Thereby a flexible and powerful method for delivering the required level of network adaptability is provided. For further details please see [6].

3.3 Adaptive Grid Control

In SmartC2Net a hierarchical approach for the control system is targeted, reflecting the structure of the components and functionalities shown in Figure 4. These are then realised for the External Generation Site as shown

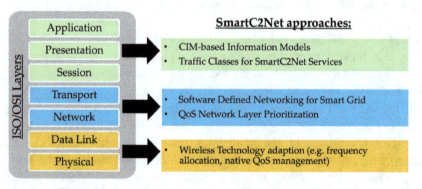

Figure 7   Network adaption and reconfiguration approaches [6]

by Figure 8 and Figure 9. The different elements serve as information sources and actuators in the control system operating over WAN and AN networks. The control platform of the project includes a Demand and Response platform, DMS functionalities for voltage control, LVGC and MVGC systems, Customer Energy Management Systems and charging controller systems, see [2] for further details.

4 External Generation Architecture Realization

In the following section we give a brief overview of the architecture and functionality of the external generation site use case.

4.1 Components and Links

The approach defined and standardized by [1] describes the relation between components and functions in terms of electrical grid components (x-axis) and zones of operation (y-axis) which is also helpful to understand the need for communication between the various components and functions.

Figure 8 illustrates those system components that will be used in order to effectively execute control of assets in the grid. For balancing the energy in

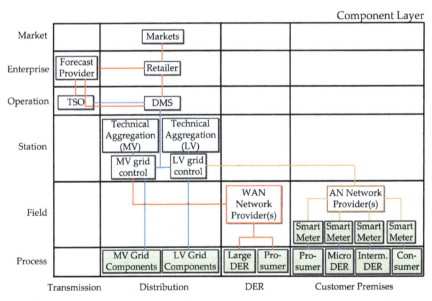

Figure 8 SGAM mapping of grid components of external generation sites [11]

the system both MV and LV grid operations are required as well as interaction with the external world as the market and TSOs. To achieve an efficient energy balancing, the control requires several different types of MV DERs, as well as having the LV grids acting as ancillary services for the MV grid operation. Since the grid is not isolated as a default, interaction with the high voltage (HV) grid is also needed as well as interaction with commercial market actors, e.g. retailers, which although shown is not in the scope of the use case.

4.2 Functional Layer

The functionalities to support the grid control are shown in Figure 9. These functionalities will be supported by the ICT platform developed in SmartC2Net. At the lowest level, actuation functionality is used to efficiently distribute actuation messages to individual assets in the grid. Protection metering and monitoring functionality is running on top of the actuation for efficient and scalable data collection and event observation. On the MV and LV side there is grid control functionality for the different voltage levels. Further on top, functionality for interaction between HV/MV and MV/LV is aggregated and controlled in coordination with any commercial aggregation,

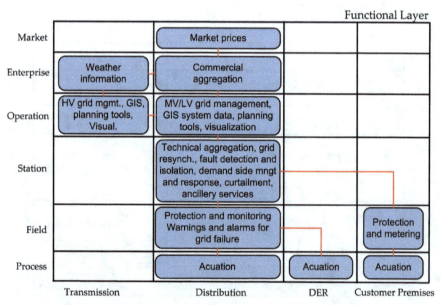

Figure 9   SGAM mapping of functionalities of external generation sites [11]

market (price) interaction. All this is coordinated with forecast providers, e.g. on the weather situation for power production prediction.

4.3 Connecting Functionalities and Components in Time

The message sequence diagram shown in Figure 10 gives an overview of the most important operation of the whole framework, focusing on the normal operation as mentioned earlier. Starting from the lowest level (right in the figure), consumers and DERs send measurements to the LV grid controller and eventually receives setpoints from the LV grid controller. The measurements from LV assets are aggregated before they are sent to the MV grid controller. In between, the LV grid controller interacts with the technical aggregator and the MV grid controller to be able to provide the setpoints to the LV assets. Similar process is ongoing for the MV grid controller, however, here interacting with the DMS and the retailers. At that level, the functionality is much focused on energy balancing, and interaction with the markets as well as forecasting services used to predict power generation from renewable energy resources. The network and the related ICT functionality will work in the background and effectively evaluates the performance of the interactions between the entities shown in Figure 10.

Thus, the overall objective of the ICT infrastructure is to ensure that the signals shown in Figure 10 are effectively mediated to the different entities involved even during and after performance degradations and faults in the network.

4.4 Requirements to the External Generation Use Case

The operation of MV and LV grids have to consider the limits for voltage variations as well as security and reliability of power supply. The set requirements to the low voltage and medium voltage grid levels are, [4, 8]:

- Voltage variations in the low voltage and medium voltage grid controllers shall not exceed 10% of nominal voltage as a 10 min. avg. value
- Voltage dips and swells shall not exceed 5% of nominal voltage (a dip/swell being defined as the difference between two samples)

For these to be fulfilled, the controllers need to be able to obtain measurements and control assets in the electrical grid. The requirements for the external generation use case has been derived in [2], and a summary of these are listed and discussed here:

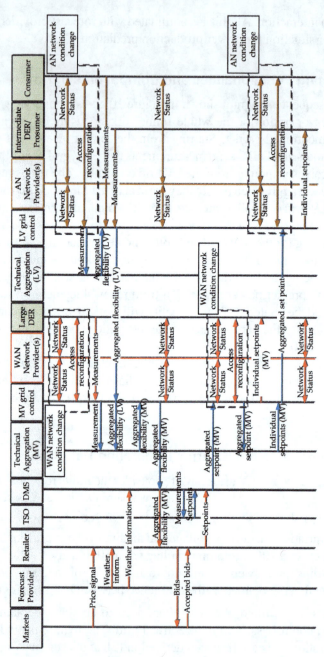

Figure 10   Message sequence diagram for interaction of the external generation site use case and adaptive network reconfiguration [11]

- Capable assets must be able to accept admissible individual setpoints for active and reactive power in order to provide proper, and as a minimum to be able to curtail.
- Assets must be able to report available active and reactive power, available defined by the difference between rated power and actual power production/consumption.
- Assets must be able to report actual reactive and active power production/consumption.
- Assets must be able to provide an operation state signal.
- Assets equipped with voltage measurements devices, shall report the voltage at the point of attachment to the electrical grid.
- Assets with energy storage capacity shall be able to provide a state-of-charge.

All these aspects relates to what is needed by the low and medium voltage controllers in order to be able to obtain the proper view of the grid state as well as be able to perform correct control actions. In addition, for the low voltage grid controller the following requirements are relevant:

- The controller shall be able to receive and follow setpoints for both active and reactive power
- The controller shall be able to calculate and dispatch setpoints for active and reactive power
- The controller shall be able to control the voltage profiles along the feeders
- The controller shall be able to aggregate and calculate the available reactive and active power within its grid domain

All these operations requires as before mentioned communication infrastructure that enables the assets, low and medium voltage controllers to be able to exchange the required information in a reliable and timely manner. Therefore, the set requirements to the communication can be summarized as:

- Performance metrics that will be used to adapt the communication and control:

1. Round trip times
2. Throughput estimates
3. Packet loss probability estimates

- Network reconfiguration interface: shall be offered to allow easy adaptation of the network while having a simple interface to the controllers

- Resilient network performance: the communication platform shall be resilient toward network faults, and be able to automatically detect network faults and mitigate faults
- Reliable middle ware components: platform components shall be reliable and robust towards faults to the degree it does not influence control performance
- Provision of meta data for accessed data: meta data regarding accessed data and distributed setpoints shall be provided

To be able to assess that these requirements are met, and that the ICT infrastructure can be considered a success, the key performance indicators already mentioned will be evaluated in a set test bed described in the subsequent section.

5 Evaluation Approach

In SmartC2Net three test beds have been defined which support the evaluation of the related use cases, however, here focusing only on the external generation site shown in Figure 11. The testbed consists of the following layers:

1. **Function and Information Layers**

- **Demand Response:** A dedicated platform for demand response is used to host functionalities related to aggregation and control of large scale flexible loads in distribution networks.
- **Primary Substation Control and Automation:** A dedicated industrial controller is used to host typical control functions in primary substations in medium voltage grids. It is also offering the possibility to implement and verify new control and operational strategies for components in medium voltage networks such as voltage control, loss minimization, etc. This industrial controller is getting information from the downstream assets placed in medium and low voltage networks.
- **Secondary Substation Control and Automation:** A dedicated industrial controller is used to host new control functionalities in secondary sub-stations (medium to low voltage). This platform offers the possibility to implement and verify new control and operational strategies for flexible assets in low voltage networks such as voltage control along the low voltage feeders, aggregation of data from smart meters, etc.
- **Plant Controller:** An industrial controller nowadays used for renewable generation plants is hosting typical control functionalities implemented in wind or PV plants.

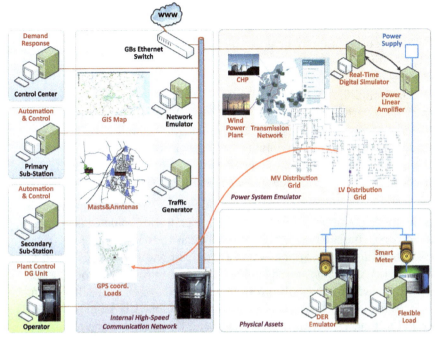

Figure 11   Architecture of the external generation site test bed

2. Communication Layer

- **Network Emulator:** A dedicated server is used to mimic the characteristics of different communication networks targeted in SmartC2net
project. For the testbed KAUNET, [7, 13] is used. Proper network models
developed in the project are realized and executed in this emulator. IP
scoping and addressing as well as the ICT platform developed is also
being executed. The servers contain also the GIS coordinates of telecom-
munication masts and any cable infrastructure available associated with
MV and LV grid.
- **Traffic Generator:** A dedicated server is used to emulate proper traffic
to add realistic, time varying traffic patterns to challenge the operation
of the grid.

3. Asset Layer

- **Dispersed Energy Resource:** A four quadrant power converter based
emulator for dispersed generation (20kW/10kVAR) is used to mimic
characteristics of small wind turbines, PV systems or energy storage.

- **House Hold:** A Controllable load of 4.5 kW is used to emulate the behavior of a typical household. A smart meter is providing power and energy consumption to the upper hierarchical control levels.
- **Real-Time Distribution Networks Simulator:** This system comprises of two elements namely the Real-Time Digital Simulator and the Power Linear Amplifier. The Real-Time Digital Simulator is implementing a realistic distribution network. The three phase voltages measured in a given point in the distribution network are applied to the Power Linear Amplifier that is supplying the physical components i.e. Dispersed Energy Resource, Household as they are part of the larger system. The three phase currents are fed in back to the Real-Time Digital Simulator.

Internally, the test bed will be supporting the IEC 61850 communication protocol where possible, for which the assets and grid emulator is attached to the network emulator via a high speed switch to focus the impact of the network emulation solely to the network emulator and not the real network. At the network emulator, several things are done: proper network models provided by other partners in the project are realized and executed, IP scoping and addressing as well as the ICT platform developed is being executed. In parallel to that, proper traffic generators are co-located in the test bed to add realistic, time varying traffic patterns to challenge the operation of the grid. For the SmartC2Net project, real life cases from different DSO's in Europe are used to create benchmark systems used to emulate the system operation and responses in real time. The key features of this Laboratory are as follows:

- **Real-Time** - simulation of distribution grids
- **Hardware-In-the-Loop** - physical industrial controllers and other components are connected to the Real-Time Digital Simulator
- **Power Hardware-In-the-Loop** - physical components that are emulating energy resources and households are connected to a distribution system as they are part of it.

6 Conclusions

In this paper the key concepts of the EU project SmartC2Net are presented targeting the support of proper smart grid control in these network environments. An overview on the required ICT architecture and its functionality is provided and a description of the External Generation Site use case is given together with the targeted evaluation approaches.

7 Acknowledgment

The research leading to these results has received funding from the European Community's Seventh Framework Programme (FP7/20072013) under grant agreement no. 318023 for the SmartC2Net project [3]. We thank all our colleges in the project for fruitful discussions and collaborations.

References

[1] CEN - CENELEC - ETSI, Smart Grids Coordination Group. *Document for the M/490 Mandate Smart Grids Reference Architecture V3.0,* 11/2012.
[2] Giovanna Dondossola, Roberta Terruggia, Sandford Bessler, Jesper Grøenbaek, Rasmus Løvenstein Olsen, Florin Iov, Christian Hägerling, Christian Wietfeld, and Davide Iacono. Smart Grid Architectures: from Use Cases to ICT Requirements. In *4th CIRED Workshop for Challenges of Implementing Active Distribution System Management,* Rome, Italy, June 2014. CIRED.
[3] European Commission. FP7 Project SmartC2Net - Smart Control of Energy Distribution Grids over Heterogeneous Communication Networks, December 2012.
[4] A. Klajn H. Markiewicz. Voltage disturbances standard en 50160-voltage characteristics in public distribution systems, 2004.
[5] Christian Hägerling et al. SmartC2Net Communication Architecture and Interfaces. *Technical Report, FP7 Project SmartC2Net,* September 2013.
[6] Christian Hägerling, Fabian Kurtz, Rasmus Løvenstein Olsen, and Christian Wietfeld. Communication Architecture for Monitoring and Control of Power Distribution Grids over Heterogeneous ICT Networks. In *Proceedings of the 3 th IEEE International Energy Conference (Energycon),* Dubrovnik, Croatia, May 2014. IEEE.
[7] Per Hurtig. Kaunet - deterministic network emulation.
[8] E. IEC. 61000-4-30: Testing and measurement techniquespower quality measurement methods, 2003. International Electrotechnical Commission Standard.
[9] Jesper-Groenbaek et al. Grid and Network Monitoring Architecture and Components, Fault Management Approach. Technical report, EU FP7 SmartC2Net project, Dec 2013.
[10] Jos-Fernn Martnez, Jess Rodrguez-Molina, Pedro Castillejo, and Rubn de Diego. Middleware architectures for the smart grid: Survey

and challenges in the foreseeable future. *Energies,* 6(7):3593–3621, 2013.

[11] Rasmus Løvenstein Olsen, Florin lov, Christian Hägerling, and Christian Wietfeld. Smart Control of Energy Distribution Grids over Hetero-geneous Communication Networks. In *accepted for publication in Proceedings of the Global Wireless Summit (GWS) 2014,* Aalborg, Denmark, May 2014.

[12] Ahmed Shawky, Rasmus L. Olsen, Jens Pedersen, and Hans-Peter Schwefel. Network Aware Dynamic Context Subscription Management. *Computer Networks,* 58(0):239–253, Jan 2014.

[13] Johan Garcia Tomas Hall, Per Hurtig and Anna Brunstrom. Performance evaluation of kaunet in physical and virtual emulation environments. Technical report, Karlstad University, Sweden, February 2012.

Biographies

Rasmus Løvenstein Olsen received his M.Sc. in 2003 from Aalborg University in the area of Intelligent Autonomous Systems at the department of Electronic Systems. He then joined the project MAGNET and MAGNET Beyond on the topic of Context Aware Service Discovery, in which he did his Ph.D. thesis work, with special focus on context management and reliability analysis of distributed, dynamic information. He then worked in several other European projects, as task and work package leader while also teaching and supervising at master and Ph.D. level. He is currently engaged in distributed information management and access in smart grids and intelligent transportation systems.

Christian Hägerling, born Müller, received his Dipl.-Ing. degree in Information Technologies from TU Dortmund University, Germany, in 2008. He is working towards his Ph.D. degree at the Communication Networks Institute (CNI), TU Dortmund University and is involved as workpackage leader in the EU FP7 project *SmartC2Net* and in the German BMWi project *E-DeMa*. His research interests focus on the analysis and performance evaluation of ICT infrastructures and wireless M2M technologies for Smart Grid applications. In this research areas he published more than 30 conference and journal papers.

Fabian Markus Kurtz received his B.Sc. and M.Sc. degrees in information technologies from TU Dortmund University, Germany, in 2011 and 2013. He is working towards his Ph.D. degree at the Communication Networks Institute (CNI), TU Dortmund University and is currently involved in EU FP7 project SmartC2Net. His research interests focus on the analysis and performance evaluation of ICT infrastructures for Smart Grid Applications.

Florin Iov (S '98, M '04, SM '06) received the Dipl. Eng. degree in electrical engineering from Brasov University, Romania, in 1993 and a PhD degree from Galati University, Romania in 2003 with a special focus in the modeling, simulation and control of large wind turbines. He was staff member at Galati University, Romania from 1993 to 2001. Dr. Iov was with Institute of Energy Technology, Aalborg University, Denmark between 2001 and 2009 where he was mainly involved in research projects regarding wind turbines and wind power systems. From 2010 to 2012 he held a position as Power System Research Specialist in Vestas Wind Systems working with new ancillary services for augmented wind power plants. Since 2013 Dr. Iov is with Institute of Energy Technology focusing on research within smart grids. His research covers control and application of electrical machines and power electronic converters for grid integration of renewable energy sources and, operation and control of dispersed generation in modern power systems. He is author or co-author of more than 110 journal and conference papers in his research areas.

Christian Wietfeld received his Dipl.-Ing. and Dr.-Ing. degrees from RWTH Aachen University. He is now a full professor and head of the Communication Networks Institute (CNI) of TU Dortmund University, Germany. For 20 years he has initiated and contributed to national and international research and development projects on wireless data communication systems in academia and industry. He has published over 135 peer-reviewed publications and holds several patents. His current research interests include the system design, modeling and performance evaluation of communication networks in challenging environments. Christian Wietfeld is a Senior Member of IEEE and chairs the IEEE's German sister organization VDE/ITG committee on Communication Networks and Systems.

Applications of Machine Learning and Service Oriented Architectures for the New Era of Smart Living

Sofoklis Kyriazakos[1], George Labropoulos[1], Nikos Zonidis[1], Magda Foti[1] and Albena Mihovska[2]

Converge ICT Solutions & Services S.A., 74, Panormou str.,
11523 Athens, Greece
CTIF, Aalborg University, Frederik Bajersvejx 7C,9220 Aalborg, Denmark

Received 28 May 2014; Accepted 28 June 2014
Publication 31 August 2014

Abstract

In this paper we present a the application of advanced pattern matching algorithms and the utilization of a cloud-based Service Oriented Architecture (SOA) to offer a number of rich personalized applications for Ambient Living, through a novel Building Management System (BMS). The novelty presented in this paper is deriving from the evolution of a proprietary BMS product, namely Ecosystem, which is enhanced with the features presented in this paper, to address high demand for personalization and service adaptation in the new era of Information and Communication Technologies.

Keywords: Ecosystem; CBR; machine learning; SOA; service elements.

1 Introduction

BMS platforms in the market today are strongly linked with industrial control hardware, meant for machinery with tight schedules, requirements, and specific operating parameters. However, people in homes and businesses have constantly varying schedules and requirements, their equipment and wiring

Journal of Communication, Navigation, Sensing and Services, Vol. 1, 181–196.
doi: 10.13052/jconasense2246-2120.124

changes all the time, and rarely can things be predicted and planned as they would be in an industrial environment.

On the same time, the new era of Information and Communication Technologies instructs non-obtrusive solutions that can provide personalized, context-aware services and applications that cannot be fully addressed by today's BMS solutions.

Ecosystem vision in contrast, conspires to create an environment that adapts to the occupants' living or working habits, and makes their life easier by providing a rich set of personalized services.

Ecosystem proprietary product is therefore included in an R&D project of the General Secretariat for Research and Technology (GSRT), co-funded by Greek and EU funds, to be enhanced with the above-mentioned features. The project has 24-months duration and is expected to result in a highly-flexible, modular and scalable platform that can be a game changer in the area of BMS solutions.

The paper is organized as follows. In Section II we present the state-of-the-art in BMS solutions and we focus on Ecosystem product, as this will be enhanced with features that are presented in this paper. In Section III, we present Machine Learning algorithms that have been designed and implemented in Ecosystem product, to allow service personalization. Furthermore, in Section IV, we analyze the SOA approach of Ecosystem, with the cloud platform and the service-element approach. In Section V the conclusions of the paper, the expected results and the future work are discussed.

2 State-of-the-art in Building Management Systems

2.1 General

A Building Management System (BMS) is a computer-based control system installed in buildings that controls and monitors the building's mechanical and electrical equipment such as ventilation, lighting, power systems, fire systems, and security systems. A BMS consists of software and hardware; the software program, usually configured in a hierarchical manner, can be proprietary, using such protocols as C-bus, Profibus, and so on. Vendors are also producing BMSs that integrate using Internet protocols and open standards such as DeviceNet, SOAP, XML, BACnet, LonWorks and Modbus. [1]

Building Management Systems are most commonly implemented in large projects with extensive mechanical, electrical, and plumbing systems. Systems linked to a BMS typically represent 40% of a building's energy usage; if lighting is included, this number approaches 70%. BMS systems are a critical component to managing energy demand. Improperly configured BMS systems are believed to account for 20% of building energy usage, or approximately 8% of total energy usage in the United States. [2, 3]

2.2 Ecosystem

Ecosystem is a cutting edge solution in the sector of Building Management Solutions. It is an affordable automation and monitoring solution which allows for extremely broad control over every aspect of a residential or commercial building. It is friendly, safe, very scalable, fault tolerant and very flexible. Like a natural ecosystem, it interacts with and adapts to its environment naturally and unobtrusively. Ecosystem reduces complications, costs, and worry for any building owner or resident.

As a pure BMS, Ecosystem is an integrated platform that is installed on the side of your current electrical and mechanical infrastructure. Ecosystem can be configured with presets, on the fly, or even by the owner of the premises, to integrate all the functionality of electrically controlled devices: sensors, lights, motors, ventilation, and so forth. Some of the automation examples are: lighting, energy saving based on sensing, computer room security, server-/UPS-management, AC/heating, video feeds and many others.

Ecosystem's automation does not mean that you lose, or sacrifice, control of your premises. In fact, quite the contrary is true: you gain much more control, as we hope you will discover below, and there is never any chance of the Ecosystem wresting control of your environment from you.

The platform is developed in Java and can manage hardware (sensors & actuators), either over a controller, or fan-less PC, or even via Internet. In terms of hardware, Ecosystem cooperates with ECHELON, KNX, Z-Wave, enocean, several PLCs, etc enabling the installation in house or industries, where twisted pair does not exist or is difficult to be installed. The software platform is very flexible and can easily manage I/Os, while it may offer services by its integration with GSM (SMS & voice in/out calls), GPRS/3G, WiFi, Bluetooth, RFID, etc. It is worth to mention that the above technologies and protocols can co-exist in an installation, as Ecosystem introduces an abstraction layer to manage sensors and actuators. Regarding the interface with the end user (in the case of home automation) this can operate either without any screen, or via mobile phone (iPhone app, html, ajax), Internet or by means of a standalone client application.

3 Machine Learning

Machine learning algorithms are data analysis methods which search data sets for patterns and characteristic structures. Over the past two decades Machine Learning has become one of the mainstays of information technology and with that, a rather central, albeit usually hidden, part of our life. With the ever increasing amounts of data becoming available there is good reason to believe that smart data analysis will become even more pervasive as a necessary ingredient for technological progress. [4]

As regards machines, we might say, very broadly, that a machine learns whenever it changes its structure, program, or data (based on its inputs or in response to external information) in such a manner that its expected future performance improves. Some of these changes, such as the addition of a record to a data base, fall comfortably within the province of other disciplines and are not necessarily better understood for being called learning. But, for example, when the performance of a speech-recognition machine improves after hearing several samples of a person's speech, we feel quite justified in that case to say that the machine has learned. [5]

Machine learning is programming computers to optimize a performance criterion using example data or past experience. We have a model defined up to some parameters, and learning is the execution of a computer program to optimize the parameters of the model using the training data or past experience. The model may be predictive to make predictions in the future, or descriptive to gain knowledge from data, or both. Machine learning uses the theory of statistics in building mathematical models, because the core task is making inference from a sample. The role of computer science is twofold: First, in training, we need efficient algorithms to solve the optimization problem, as well as to store and process the massive amount of data we generally have. Second, once a model is learned, its representation and algorithmic solution for inference needs to be efficient as well. In certain applications, the efficiency of the learning or inference algorithm, namely, its space and time complexity, may be as important as its predictive accuracy. [6]

3.1 Types of Learning

Machine learning algorithms can be organized into a taxonomy based on the desired outcome of the algorithm or the type of input available during training the machine.

Supervised learning algorithms are trained on labeled examples, i.e., input where the desired output is known. The supervised learning algorithm attempts to generalize a function or mapping from inputs to outputs which can then be used to speculatively generate an output for previously unseen inputs.

Unsupervised learning algorithms operate on unlabeled examples, i.e., input where the desired output is unknown. Here the objective is to discover structure in the data (e.g. through a cluster analysis), not to generalize a mapping from inputs to outputs. [7]

3.1.1 Supervised Learning

Supervised learning entails learning a mapping between a set of input variables Ξ and an output variable Y and applying this mapping to predict the outputs for unseen data. Supervised learning is the most important methodology in machine learning and it also has a central importance in the processing of multimedia data. [8]

In supervised learning, we know (sometimes only approximately) the values of f for the m samples in the training set, Ξ as shown in Figure 1. We assume that if we can find a hypothesis, h, that closely agrees with f for the members of Ξ, then this hypothesis will be a good guess for f — especially if Ξ is large.

Training Set:

$$\Xi = \{X_1, X_2, \ldots X_i, \ldots, X_m\}$$

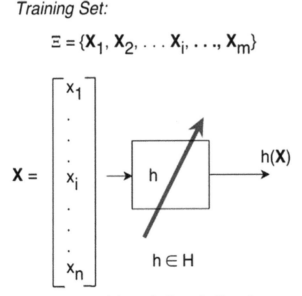

Figure 1   Training set for Supervised Learning

We will use x(i) to denote the "input" variables, also called input features, and y(i) to denote the "output" or target variable that we are trying to predict. A pair (x(i), y(i)) is called a training example, and the dataset that we'll be using to learn—a list of m training examples {(x(i), y(i)); i = 1, . . . , m} is called a training set. To describe the supervised learning problem slightly more formally, our goal is, given a training set, to learn a function h : X → Y so that h(x) is a "good" predictor for the corresponding value of y. For historical reasons, this function h is called a hypothesis. Seen pictorially, the process is therefore shown in Figure 2.

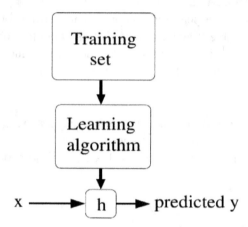

Figure 2   Learning process through training

When the target variable that we're trying to predict is continuous we call the learning problem a regression problem. When y can take on only a small number of discrete values (such as if, given the living area, we wanted to predict if a dwelling is a house or an apartment, say), we call it a classification problem.

3.1.2 Unsupervised Learning

In unsupervised learning, we simply have a training set of vectors without function values for them. The problem in this case, typically, is to partition the training set into subsets, $\Xi1, . . . , \Xi R$, in some appropriate way. (We can still regard the problem as one of learning a function; the value of the function is the name of the subset to which an input vector belongs.) Unsupervised learning methods have application in taxonomic problems in which it is desired to invent ways to classify data into meaningful categories. [5]

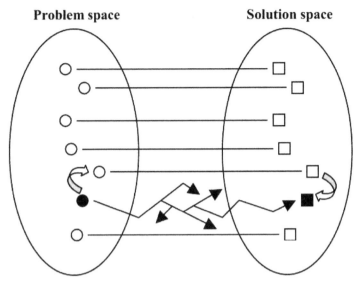

Figure 3 Problem solving in CBR

3.2 Ecosystem

Ecosystem uses Case Based Reasoning by implementing a supervised lazy problem-solving method. Case-Based Reasoning (CBR) is an approach to problem solving that emphasizes the role of prior experience during future problem solving, CBR solves new problems by adapting previously successful solutions to similar problems. [9]

Figure 3 illustrates how the assumptions listed above are used to solve problems in CBR. Once the currently encountered problem is described in terms of previously solved problems, the most similar solved problem can be found. The solution to this problem might be directly applicable to the current problem but, usually, some adaptation is required. The adaptation will be based upon the differences between the current problem and the problem that served to retrieve the solution. Once the solution to the new problem has been verified as correct, a link between it and the description of the problem will be created and this additional problem-solution pair (case) will be used to solve new problems in the future. Adding of new cases will improve results of a CBR system by filling the problem space more densely. [10]

According to Kolodner, the CBR working cycle can be described best in terms of four processing stages:

- **Case retrieval**: after the problem situation has been assessed, the best matching case is searched in the case base and an approximate solution is retrieved.
- **Case adaptation**: the retrieved solution is adapted to fit better the new problem.
- **Solution evaluation**: the adapted solution can be evaluated either before the solution is applied to the problem or after. In any case, if the accomplished result is not satisfactory, the retrieved solution must be adapted again or more cases should be retrieved.
- **Case-base updating**: If the solution was verified as correct, the new case may be added to the case base.

Aamodt and Plaza [11] give a slightly different scheme of the CBR working cycle comprising the four REs [12]:

- **RETRIEVE** the most similar case(s);
- **REUSE** the case(s) to attempt to solve the current problem;
- **REVISE** the proposed solution if necessary;
- **RETAIN** the new solution as a part of a new case.

Ecosystem uses the lazy method of Case Based Reasoning because CBR has the following characteristic:

- Easy and comprehensible representation of cases
- Scalability of the case base
- Easy acquisition and upgrading of knowledge while system is operating
- Efficiency. It is easier to adjust a solution given in an earlier problem than solve the problem from the beginning.

3.3 Case Representation

In Ecosystem a case comprises a:

Problem description, which represents the state of the world when the case occurred. More specifically, the problem description consists of a feature vector which carries the values of the sensors when the problem occurred and the date or time of the day during which the case is active.

Problem solution, which states the derived solution to that problem. The solution is a feature vector. Each feature consists of three parts, the identifier of the actuator, the action to be taken and a value assigned to each feature after the evaluation of the case, which takes place after the implementation of the case.

3.3.1 Indexing

The indexing of the cases is based on features of the case that have the information of the timeframe during which the case is active. This time frame may appear in two forms, it may include a period of days for each year or a period of time during each day. Each case has been assigned a composite label that describes the time of day and the time of year this case is active. In this way indexing method is predictive, recognizable and discriminating enough to facilitate efficient and accurate retrieval.

3.3.2 Case base organization

Ecosystem uses a flat organization for its case base. Flat organization is the simplest case-base organization that yields a straightforward flat structure of the case base. The advantages of this organization is that it is simple, uses a clear flat structure for the case base and addition/deletion of cases is done easily without computational cost.

3.3.3 Retrieval

Ecosystem uses 1st-NN algorithm for case retrieval. Nearest neighbor techniques are perhaps the most widely used technology in CBR since it is provided by the majority of CBR tools [13]. Nearest neighbor algorithms all work in a similar fashion. The similarity of the problem (target) case to a case in the case-library for each case attribute is determined. This measure may be multiplied by a weighting factor. Then the sum of the similarity of all attributes is calculated to provide a measure of the similarity of that case in the library to the target case. This can be represented by the equation:

$$Similarity\,(T, S) = \sum_{i=1}^{n} f(T_i, S_i) \times w_i \qquad (1)$$

where T is the target case; S the source case; n the number of attributes in each case; i an individual attribute from 1 to n; f a similarity function for attribute i in cases T and S; and w the importance weighting of attribute i [14].

3.3.4 Revise

After the retrieval and the implementation of a case Ecosystem evaluates the solution. Ecosystem follows a type of structural adaptation of the stored case.

Each solution as described above consists of a feature vector. Each feature represents an action to be undertaken by our system. During the evaluation phase, two of the three parts of the solutions can be changed, these are the action to be performed and the evaluation score. The rating of each characteristic varies depending on the actions the user takes after the implementation of the solution. If for example, the solution recovered includes, as one of its features, the activation of lighting in a point of the house and the user proceeds to the immediate deactivation of this feature takes negative rating. The features that collect maximum points allowed by the system, are eliminated from the stored solution.

A specificity of the Ecosystem is that its CBR system does not make case additions to its case base after revisions, but an evalutation an adaptation of the existing cases. that has not made a case adding to base his case, but an evaluation and adaptation of existing cases. Addition of cases is made only by the user, who initializes the system by inserting cases and who can add or delete cases at any time.

4 Cloud BMS

The notion of administering Ecosystem instances creates a pivotal point which pushes Ecosystem into the current developments related to cloud computing. Central administration of several instances of Ecosystem installations is supported by a central web application which connects privately and safely to every instance allowing for users to access and control each or a number of them according to their privileges. A number of specific services published by each Ecosystem, if configured to do so are subscribed to the web server. From there on these services provide data both incoming and outgoing, therefore providing readings of Ecosystem metrics as well as enforcement of specific actions on the actuators orchestrated by a particular Ecosystem instance. Therefore, two major pillars support the Cloud BMS functionality:

- Ecosystem Service elements
- Central administration web server

The service elements published by Ecosystem support the communication with and remote control of it by the user who is permitted to do so. They are uniquely accessible for each instance of Ecosystem through the relevant URL, e.g.

https://ecosystem1.hostname1:5001/getAvailableServices

This for example provides a list with all the available services of the particular Ecosystem. Once combined, many services together can result in the remote orchestration and management of the instance or the gathering of a particular reading (e.g. Temperature) from multiple instances that will provide more reliable input for a particular process required by an Ecosystem installation. An example of how the Services are used is shown in the diagram Figure 4.

Apart from the remote user who has access to the Ecosystems through the web application, other systems can have access two by calling essentially the services of the Ecosystems through the web server. A typical use of such a feature is the gathering of services into a process that executes a particular scenario by using of ESB software such as MuleSoft [15]. Processes designed with MuleSoft ESB can be executed as standalone web applications. In these processes, several service elements published from the central administration Web server are connected on the ESB dashboard according to the logic that needs to be created. Once a process is formulated it can be tested and deployed as a web application. The purpose of combining service elements boosts the performance of Ecosystem as the unrelated services which correspond to sensors and actuators are combined in order to provide more complex computations and decision making. The image Figure 5 displays an example process designed using the MuleSoft SDK. The building blocks are common actions varying from standardized events such as HTTP/HTTPS to SAP connections and AJAX calls. These components are customized by the user once placed on the dashboard, so in our case, several HTTP calls made to the central admin web server combined together provide us with several inputs to decide an action that must be performed on a particular Ecosystem.

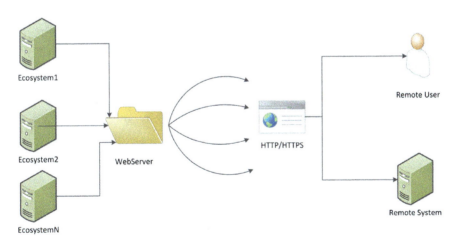

Figure 4 Service availability from various ecossytem instances

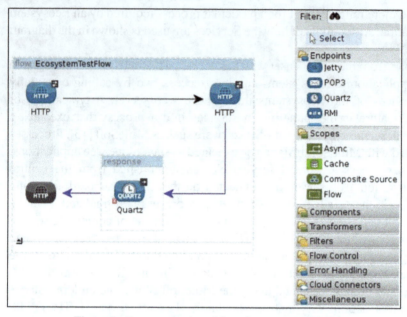

Figure 5   Processes and workflows in ecosystem BMS

5 Conclusions

In this paper we have focused on the limitations of today's BMS solutions and we presented means to upgrade an existing proprietary BMS solution, namely Ecosystem from Converge. We expect that the use of machine learning mechanisms and the development the cloud BMS component for Ecosystem, will result in a highly flexible BMS system that will be a game changer in the sector. In addition, it will allow a large number of services and applications to be developed and deployed; thus extending its lifecycle by involving developers from the community interested to exploit the Ecosystem platform. As future steps we will validate the platform in a real environment and through scenarios and use-cases addressing mainly eHealth applications and Ambient Assisted Living. It is expected that the validation will start in less than one year and results will be available a few months later.

6 Acknowledgment

Part of this work has been carried out in the scope of the ECOSYSTEM project, which is a General Secretariat for Research and Technology (GSRT) project, co-funded by Greek and EU funds. The authors acknowledge help and contributions from all the partners of the project.

References

[1] Wikipedia, Building management system, http://en.wikipedia.org/wiki/Building_management_system

[2] Advanced Sensors and Controls for Building Applications: Market Assessment and Potential R&D Pathways (Brampley 2005)

[3] Energy Consumption Characteristics of Commercial Building HVAC SystemsVolume III: Energy Savings Potential (Roth 2002)

[4] A. Smola and S. Vishwanathan, Introduction to Machine Learning, Cambridge University Press, 2008.

[5] N. Nilsson, Introduction To Machine Learning, Stanford University, 1998.

[6] E. Alpaydim, Introduction To Machine Learning, The MIT Press, 2010.

[7] http://en.wikipedia.org/wiki/Machine_learning.

[8] M. Cord and P. Cunningham, Machine Learing Techniques for Multimedia Case Studies on Organization Retrieval, Springer, 2008.

[9] R. L. Mantaras, D. Mcsherry, D. Bridge, D. Leake, B. Smyth, S. Craw, B. Faltings, M. L. Maher, M. Cox, K. Forbus, M. Keane, A. Aamodt and I. Watson, "Retrieval, reuse, revision and retention in case-based reasoning," The Knowledge Engineering Review, vol. 20, no. 3, pp. 215–240, 2006.

[10] M. Pantic, Introduction to Machine Learning & Case-Based Reasoning, London: Imperial College

[11] A. Aamodt and E. Plaza, "CBR: foundational issues, methodological variations and system approaches," AI Communications, vol. 7, no. 1, pp. 39–59, 1994.

[12] I. Watson and F. Marir, Case-Based Reasoning: A Review, New Zealand: University of Auckland, 2000.

[13] I. Watson, Applying Case-based Reasoning: Techniques for Enterprise Systems, USA: Morgan Kaufmann, 1997.

[14] I. Watson, "Case-based reasoning is a methodology not a technology," Knowledge-Based Systems, vol. 12, pp. 303–308, 1999.

[15] http://www.mulesoft.org/

Biographies

Dr. Sofoklis Kyriazakos graduated Athens College school in 1993 and obtained his Master's degree in Electrical Engineering and Telecommunications in RWTH Aachen, Germany in 1999. Then he moved to the National Technical University of Athens, where he obtained his Ph.D. in Telecommunications in 2003. He also received an MBA degree in Techno-economic systems from the same university. He has more than 90 publications in international conferences, journals, books and standardization bodies. He has been invited as reviewer, chairman, member of the committee, panelist and speaker in many conferences and has also served as TPC chair in 2 International conferences. Currently he holds the academic position of Associate Professor in the University of Aalborg. Sofoklis has managed, both as technical manager and coordinator, a large number of multi-million ICT projects, at R&D and industrial level. In 2006 Sofoklis founded an ICT startup, Converge S.A., with the PRC Group and since then he has been the Managing Director and a BoD member. Since March 2012, he also Managing Director and BoD member of Converge ICT Innovation Inc. based in Montreal, Canada, which is an affiliate company. Sofoklis is also member of the BoD of Athens Information Technology, a Center of Excellence for Research and Education.

George Labropoulos is the CTO of Converge since February 2012 and is an IT Specialist with large experience in large corporations and has also worked as

IT Consultant for many years. He studied Informatics and Computer Science in Athens University and York University in Toronto. He obtained his degree in Computer Programming and & Systems with Honors from Devry Institute of Technology in Toronto.

Nikolaos Zonidis was born in Athens on 12th March 1980. He holds a Bsc on Computer and Network Techologies from Northumbria University at Newcastle UK after which he completed his Msc on Computer and Internet Technologies at Glasgow Strathclyde University, focusing on client-server systems and software implementation. His thesis was based on sandboxing of suspicious software on the context of personal computing. since Jue 2006 he is active in JavaEE development for several business models in particular telecommunications billing and sales channels. His research activities include IoT and M2M communication protocols.

Magda Foti was born in Naoussa, Greece, in June 1985. She received an Engineer's degree and a Master degree from University of Thessaly, Department of electrical and computer engineering, in 2009 and 2011 respectively. Since 2011 she works as a Software and R&D Engineer at Converge ICT Solutions & Services S.A. She was involved in various R&D projects in the area of IoT and machine learning. Her main research interests are in the broad areas of machine learning, natural language processing and decision theory.

Albena Mihovska has a PhD degree in mobile communications from Aalborg University, Aalborg, Denmark, where she is currently an Associate Professor and Head of Standardization and Head of Teaching at the Center for TeleInfrastruktur (CTIF). Currently, she is involved with research related to innovative research concepts for 5G communication systems, the design and implementation of e-Health services (EU project eWALL) and to optimizing and supporting reliable and high performance intensive data rate communications as required by the Internet of Things.

www.ingramcontent.com/pod-product-compliance
Lightning Source LLC
LaVergne TN
LVHW012332060326
832902LV00011B/1858